Journey into Black Holes

OrangeBooks Publication

1st Floor, Rajhans Arcade, Mall Road, Kohka, Bhilai, Chhattisgarh 490020

Website: **www. orangebooks. in**

First Edition, 2026

ISBN: 978-93-7426-576-5

JOURNEY INTO BLACK HOLES

DR. SHARAD CHANDRA BHAIYA

OrangeBooks Publication
www.orangebooks.in

Dedicated to all those who look at the night sky and ask "Why"?

Journey into Black holes

Discovery, Detection, Hidden Secrets, and Recent Discoveries

- Dr. Sharad Chandra Bhaiya

For every curious mind — exploring the universe is simple.

Preface

This book presents a comprehensive journey through the fascinating world of black holes—from their early theoretical predictions and discovery to their formation, detection methods, and classification. It further explores binary and triple systems, the physical characteristics of black holes, and extreme gravitational phenomena such as gravitational redshift, time dilation, and spaghettification. Special attention is also given to rogue black holes and notable examples of unique or extraordinary systems.

The book highlights recent groundbreaking discoveries made by the LIGO, Virgo, and KAGRA observatories, along with contributions from several other international collaborations. These achievements have revolutionized our understanding of black hole mergers and gravitational waves, providing the first direct observational evidence of these invisible giants. Recent findings—such as the triple systems V404 Cygni and D9, Triple AGN system and the discovery of a runaway supermassive black hole drifting through intergalactic space—underscore the rapidly evolving nature of modern astrophysics.

The primary aim of this book is to make the complex and mysterious subject of black holes accessible to general readers who are curious about the universe but may not have a scientific background. Every effort has been made to keep the language clear, engaging, and easy to understand, while maintaining scientific accuracy. *As Albert Einstein wisely said, "If you can't explain it simply, you don't understand it well enough,"* and this principle has guided every explanation in this book.

By blending scientific facts, modern discoveries, and poetic reflections, this book seeks to spark curiosity, inspire wonder, and make the invisible world of black holes tangible for all readers. It is an invitation to explore the cosmos not merely as a distant observer, but as an active participant in humanity's quest to understand the universe. A selection of authentic and visually striking images has also been included to enrich the reader's understanding of these extraordinary celestial objects.

The meanings of certain uncommon terms and important scientific expressions used throughout the book are explained in the Glossary and Vocabulary sections. The meanings and full forms of astronomical names of systems and cosmic objects are provided in Appendix B: **Understanding Astronomical Names.** Readers are encouraged to refer to these sections whenever they encounter unfamiliar words or concepts, making the text easier to understand and more enjoyable to read.

Author kindly requests readers to excuse any inadvertent errors that may appear in this book. Constructive suggestions and thoughtful feedback are most welcome and will be gratefully considered for future editions.

A second volume—focusing on observatories dedicated to black hole discovery and research, along with biographies of eminent scientists and astronomers—is planned for future publication.

Acknowledgment

The author expresses sincere gratitude to the scientists, researchers, and institutions whose open-access publications, observational data, and public outreach efforts have significantly contributed to the creation of this book. Special acknowledgment is extended to the global gravitational-wave community, including the teams of the LIGO Scientific Collaboration, Virgo Collaboration, and KAGRA Collaboration, whose groundbreaking discoveries have revolutionized our understanding of black holes and the dynamic universe.

Information has been adapted from publicly available scientific sources, including NASA, European Space Agency (ESA), the European Southern Observatory (ESO), the Event Horizon Telescope Collaboration, and other reliable scientific platforms such as Wikipedia.

Special appreciation is extended to the many astronomers, engineers, mission teams, and observatories whose discoveries continue to expand our understanding of the universe and inspire scientific curiosity.

All scientific explanations in this book have been independently rewritten for educational and outreach purposes. No proprietary material has been copied.

Journey into Black Holes

Discovery, Detection, Hidden Secrets, and Recent Discoveries

- Dr. Sharad Chandra Bhaiya

Basic Astronomical Units and Terms

Solar Mass ($M_\odot$): The solar mass is the mass of our Sun and is used as a standard unit to measure the masses of stars and black holes.

1solar mass $\mathbf{M_\odot} \approx 2\times10^{30}$ kg

Light-Year (ly):

A light-year is the distance that light travels in one year (365.25 days) in empty space.

1light year(ly) $\approx 9.46\times 10^{12}$ km

Astronomical Unit (AU): It is the mean distance from the Earth to the Sun, about 150 million kilometers. It serves as a basic measuring stick for distances inside the Solar System.

1AU $\approx 1.496\times10^{8}$ km

1 light-hour: The distance travelled by light in one hour

1light- hour $\approx$ 1.08 billion kilometers

Velocity of light: The distance travelled in vacuum by light in 1 sec

Velocity of light **(c)** $\approx 3.0\times10^{8}$ m/s

Parsec (pc): A unit of astronomical distance defined as the distance at which one astronomical unit subtends an angle of one arcsecond; it equals about 3.26 light-years or 3.086×10^{13} km.

Table of Contents

Chapter 1

The Discovery and Detection of Black Holes

Introduction

Despite being invisible, inaccessible, and impossible to observe directly, black holes remain among the most fascinating and captivating objects in the universe. These mysterious entities—stranger than fiction—have intrigued scientists and the public alike for more than a century. Over the past hundred years, a panorama of groundbreaking scientific discoveries has unfolded, and among the most profound of these is the realization that black holes are real physical objects.

Black holes have long attracted attention not only from scientists but also from society as a whole—out of curiosity, fascination, and sometimes fear. Their extraordinary properties challenge our understanding of nature. A black hole can trap anything—from massless photons to massive stars—within a remarkably small region of spacetime drawing them inward on a very short timescale once the boundary is crossed.

Once considered purely theoretical constructs, black holes are now firmly established as real cosmic entities. This transition from speculation to confirmation has been driven by compelling observational evidence and landmark discoveries. Today, black holes are among the most intensely studied objects in astrophysics, yet they remain far from fully understood.

For theorists, black holes offer a natural laboratory for exploring the deepest laws of physics under extreme conditions. Before black holes were ever directly observed or confirmed, they had already been carefully classified in theory. Scientists were so confident in the soundness of Einstein's equations and gravitational physics that they were certain these mysterious and extreme objects must exist in nature. Their properties, types, and behavior were described mathematically long before telescopes and detectors revealed their presence in the cosmos. This remarkable fact demonstrates the power of theoretical science: prediction came first, and discovery followed.

For experimentalists, however, black holes are far more difficult to study, because they neither emit light nor reflect any light that falls upon them. This complete darkness makes them invisible against the cosmic background and extremely challenging to locate directly.

In addition, black holes are often embedded in dense and turbulent environments, such as the centers of galaxies or regions of intense star formation, where dust and gas obscure observations. Their immense distances from Earth further weaken observable signals, and many black holes remain dormant, showing little or no accretion activity for long periods. All these factors together make black holes some of the most elusive objects in the Universe, turning their study into a remarkable scientific challenge that continues to push the limits of observational and experimental techniques.

The Birth of the Black Hole Idea

The earliest glimpse of a black hole can be traced back to 1784, when British **philosopher, geologist,** and **astronomer John Michell (1724-893)** proposed a revolutionary idea. He suggested that a sufficiently massive and compact star could possess an escape velocity greater than the speed of light, preventing light from escaping its surface and rendering the object invisible.

Michell's insight was remarkable for its time and laid the conceptual groundwork for black hole physics. He is often regarded as one of the earliest pioneers in the field.

Michell was a unique figure—a clergyman by profession and an astronomer by passion—who spent his days as a priest and his nights exploring the mysteries of the cosmos. He is recognized as a **father of black hole physics.**

In 1796, French mathematician, astronomer and physicists **Pierre-Simon Laplace** independently arrived at a similar conclusion. He proposed that a star of sufficient mass, size, and density could become invisible and referred to such an object as a *"**dark star.**"* Laplace's treatment was mathematical in nature and provided a formal theoretical framework for the idea.

Figure 1.1: Pierre - Simon Laplace (1749-1827)

Image credit: Public domain

For more than a century, the concept of dark stars remained largely dormant. In **1915,** Albert Einstein's **theory of general relativity** rekindled interest by providing the fundamental theoretical foundation for understanding extremely compact objects. Both the expanding universe and black holes emerge naturally from this theory.

General Relativity and Black Holes

The theory of general relativity, proposed by **Albert Einstein** in November **1915,** provides the fundamental framework for understanding how black holes form from massive stars. According to this theory, gravity is not a force in the traditional sense but a **consequence of the curvature of spacetime caused by mass and energy.** The more massive an object is, the more strongly it bends the surrounding spacetime.

Figure 1.2: Albert Einstein (1879-1955)

Image credit: Public domain

When a very massive star exhausts its nuclear fuel, it can no longer support itself against its own gravity. If the remaining core is sufficiently massive, gravity dominates all known forces and the collapse continues without any known mechanism to halt it. General relativity predicts that such an extreme collapse would produce an object so compact and dense that its gravitational pull becomes extraordinarily strong, trapping all matter and radiation within a finite region of space. From the outside, this object would appear completely dark and isolated, influencing its surroundings only through its intense gravitational field.

At the time these ideas were proposed, neither the modern terminology nor the full physical interpretation existed. Scientists referred to it as a "**collapsed star**," a "**gravitationally frozen object,**" or a region of spacetime from which no signals could escape, laying the conceptual foundation for what would later be formally identified and named.

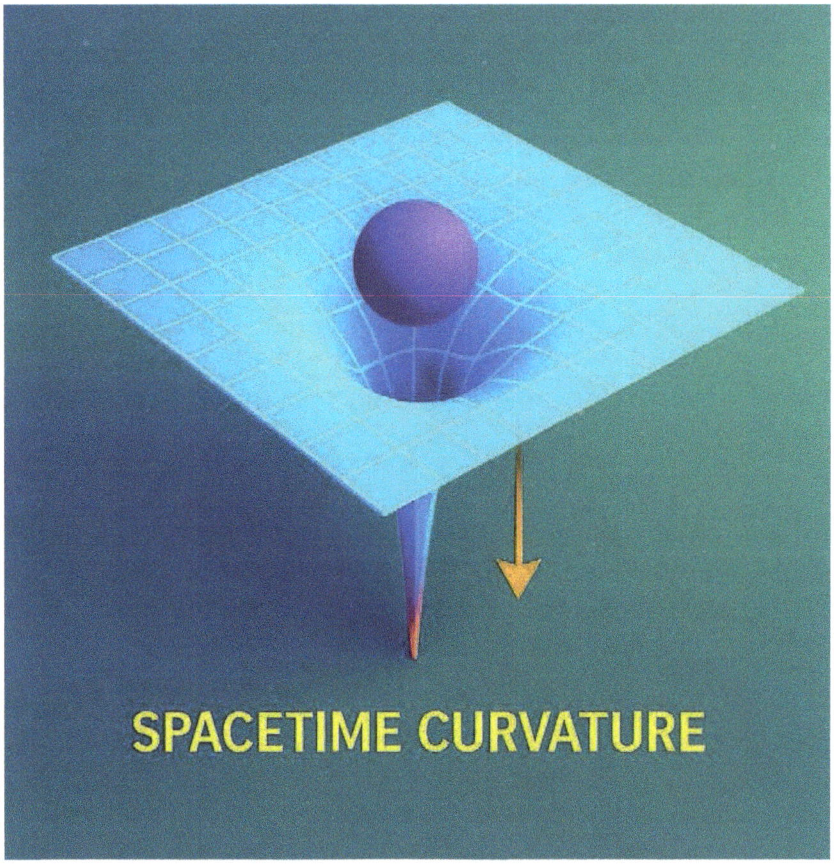

Figure 1.3: AI-generated Illustration of Einstein's spacetime curvature: mass distorts a spacetime, producing the effect of gravity for conceptual purposes

Einstein's equations allow exact mathematical solutions describing how spacetime behaves around such collapsed objects, showing how matter and energy determine the geometry of spacetime. Thus, general relativity not only explains gravity itself but also provides the theoretical link between the life cycle of massive stars and the formation of black holes.

Although Albert Einstein developed the theory that predicted the existence of what we now call of black holes, he himself was not convinced. In 1939, he published a paper expressing doubts that such rare objects could form in nature. He believed that internal forces within stars would prevent complete gravitational collapse.

Mathematical Solutions of Black Holes

In 1916, only months after Albert Einstein published his theory of general relativity, the German astrophysicist **Karl Schwarzschild** found the first exact solution to Einstein's field equations. His solution described the spacetime geometry surrounding a perfectly spherical, non-rotating mass. It revealed the existence of a **critical radius**—now known as the **"Schwarzschild radius"**—within which the curvature of spacetime becomes extreme. Although Schwarzschild himself did not interpret this result as a physical object like a black hole, later scientists recognized that this solution describes what we now call a black hole, where gravity becomes so strong that nothing, not even light, can escape.

Figure 1.4: Karl Schwarzschild (1873-1916), German astronomer and physicists

Image credit: Public domain

Despite the mathematical validity of Schwarzschild's solution, it was initially met with deep skepticism. Many scientists believed that unknown physical processes would intervene and prevent nature from producing such extreme and exotic objects. The idea that gravity alone could compress matter to such extraordinary densities seemed almost unimaginable.

In the **1930, Subrahmanyan Chandrasekhar,** the Indian-American astrophysicist, made a decisive breakthrough by applying quantum mechanics and special relativity to stellar structure. His work revealed that stars do not all share the same fate; instead, their final state is determined primarily by their mass. This insight laid the foundation of the modern theory of stellar evolution and provided the first systematic framework for understanding compact stellar remnants.

Figure 1.5: Subrahmanyan Chandrasekhar (1910-1995)

Image credit: NASA Starchild Project, Public domain

Theory of Stellar Evolution: From Star to White Dwarf

Although Einstein's general theory of relativity and Schwarzschild's solution had shown that spacetime could become extremely curved, there was still no clear physical theory explaining which stars could actually collapse into the most compact possible objects. Chandrasekhar's theory of stellar evolution filled this gap. It showed how ordinary stars could naturally evolve into compact objects through well-defined physical processes.

According to his theory, stars with initial masses not more than about **eight times the mass of the Sun**, after exhausting their nuclear fuel, expel their outer layers and leave behind a hot, dense core. This remnant is supported by **electron degeneracy pressure** and is known as a "**white dwarf.**"

However, Chandrasekhar demonstrated that a white dwarf can remain stable only if its mass is below a critical value of about **1.44 M⊙,** now called the **Chandrasekhar limit.** Beyond this limit, electron degeneracy pressure is no longer sufficient to resist gravity.

He showed that stars whose cores exceed this critical mass cannot end their lives as white dwarfs. Once their nuclear fuel is exhausted, they must undergo further gravitational collapse, forming objects even more compact than white dwarfs. In this way, Chandrasekhar's work opened the theoretical pathway to the existence of denser stellar remnants, more massive than a white dwarf.

Although white dwarfs had already been observed, they were long regarded as strange anomalies. The Chandrasekhar limit provided a solid theoretical foundation explaining both their existence and their stability. It transformed white dwarfs from curiosities into a natural and inevitable outcome of stellar evolution.

Chandrasekhar also firmly supported the idea that nature could indeed produce objects of extraordinary compactness, as suggested by Einstein's and Schwarzschild's equations. He accepted that gravity might compress matter so intensely that an object could become so dense and so strongly curved in spacetime that even light would be unable to escape. His work thus formed the bridge between stellar evolution theory and the later understanding of the most extreme gravitational objects in the Universe.

Although Chandrasekhar's conclusions were mathematically sound, they were strongly criticized at the time. During a presentation in Cambridge in 1935, his ideas were openly challenged by the eminent astrophysicist **Sir Arthur Eddington,** who argued that some unknown physical mechanism would ultimately prevent such complete gravitational collapse. This opposition delayed the acceptance of Chandrasekhar's work, even though it was firmly based on well-established physical principles.

From Star to Neutron Star

Encouraged by the Chandrasekhar limit and by the **discovery of the neutron** by **James Chadwick** in **1932,** astronomers **Walter Baade** and **Fritz Zwicky** proposed in **1934** that more massive stars (more than 8 solar masses) could leave behind extremely dense remnants composed almost entirely of neutrons. They proposed that during a supernova explosion, intense gravitational forces cause electrons and protons to combine, forming neutrons and giving rise to a new and extraordinary type of celestial object: the **neutron star**. This object is far denser and more compact than a white dwarf. Just as electron degeneracy pressure supports a white dwarf against gravitational collapse, they proposed that **neutron degeneracy pressure** could support these neutron-rich remnants.

Later theoretical developments further refined that stability in neutron stars is provided not only by neutron degeneracy pressure, but also by the repulsive component of the strong nuclear force acting between neutrons at very short distances. Together, these effects counterbalance gravity and allow neutron stars to exist as stable compact objects.

The Limits of Neutron Star Stability: The TOV Limit

In 1939, Tolman, Oppenheimer, and **Volkoff** developed a relativistic theory describing the internal structure of neutron stars and established the concept of a maximum mass for such objects, now known as **the Tolman–Oppenheimer–Volkoff (TOV) limit**. They showed that beyond this limit, no known force can prevent further collapse, leading inevitably toward a black hole. Their work demonstrated that neutron stars themselves have an upper mass limit beyond which they cannot remain stable.

Discovery of Neutron Stars

Confirming Through Observations

Despite this strong theoretical foundation, neutron stars remained purely hypothetical for more than three decades. In **1967, Jocelyn Bell Burnell**, a graduate student at Cambridge University, made a historic discovery.

Figure 1.6: Jocelyn Bell Burnell (1943 – present)

Image credit: Wikimedia Commons (CC01.0)

She detected rapidly repeating radio signals coming from a compact celestial source. These pulses were extraordinarily regular, indicating a highly stable and rapidly rotating object.

Further study showed that the source was extremely small, incredibly dense, and had a mass greater than the Chandrasekhar limit for white dwarfs but less than about three times the mass of the Sun. This matched precisely the predicted properties of neutron stars. The object was soon understood to be a rotating neutron star emitting beams of radio radiation. As it spins, these beams sweep across space like a lighthouse, producing the observed pulses.

Such objects are now called **pulsars**. The discovery of pulsars provided the first direct observational proof of neutron stars and confirmed one of the most remarkable predictions of theoretical astrophysics. At that time, neutron stars were recognized as the most compact stable objects known in the Universe.

Theoretical Foundations of Black Holes

From Star to Black Hole

Soon after the theoretical development of neutron stars, in 1939 physicists **J. Robert Oppenheimer** and his student **Hartland Snyder** published a landmark paper showing that a sufficiently massive star—typically one born with more than about **20 times the mass** of the **Sun,** after exhausting its nuclear fuel, would undergo unstoppable gravitational collapse. Using Einstein's general theory of relativity, they demonstrated that if no known force can counter gravity, the collapsing core must continue to shrink indefinitely.

From the point of view of a distant observer, the star would appear to become increasingly dim and redshifted, gradually fading from view. Although the term black hole had not yet been coined, their work provided the first rigorous and complete theoretical description of what we now recognize as black hole formation.

Later theoretical developments showed that the final outcome of stellar collapse depends critically on the mass of the stellar core. If the collapsing core exceeds roughly **three times the mass of the Sun**—surpassing the maximum mass that neutron degeneracy pressure can support—no known force can halt the gravitational collapse. The contraction then proceeds inevitably, resulting in the formation of an object even more compact and dense than a neutron star: a black hole. They also concluded that to produce such a massive collapsing core, the original star must itself be very massive, typically having an initial mass greater than about **20 solar masses**. In this way, Oppenheimer and Snyder provided the theoretical gateway to the existence of the most extreme compact objects in the Universe.

Their work clearly established the minimum conditions required for collapse beyond the neutron star stage and showed that nature allows the formation of objects even more compact and gravitationally intense than neutron stars. This marked the birth of black holes as a solid prediction of physical theory, long before their observational discovery.

From Theory to Reality: The Discovery of Black Holes

The strong theoretical prediction of black holes, together with the experimental discovery of neutron stars, sparked intense interest in the astronomical community and motivated the search for observational evidence of more compact objects greater than neutron stars. Scientists realized that even more extreme collapsed objects must also exist. This realization led astronomers to look for indirect signatures of black holes.

But black holes neither emit nor reflect light, making them invisible to ordinary telescopes. For many years, this posed a major challenge for astronomers, as there were no reliable observational tools to locate them. Often hidden in distant regions of space, black holes appeared to leave no direct trace of their presence.

However, astronomers eventually realised that black holes reveal themselves through their immense **gravitational influence on surrounding matter**.

When a black hole pulls in gas and dust from a nearby star or its environment, the infalling material generates intense friction and heat. Extreme gravitational forces heat this matter to **millions of degrees**, causing it to emit **strong X-rays**. These X-rays act as a distinctive signature, allowing astronomers to detect otherwise invisible black holes.

Early Detection of Cosmic X-ray Sources

The search for cosmic X-ray sources began with rocket- and balloon-based experiments in the early **1960s.** In **1962,** sounding rocket observations led to the discovery of **Scorpius X-1, the first known extrasolar source of X-rays,** marking the birth of X-ray astronomy. During the same period, rocket experiments also detected several other bright X-ray sources, including the Cygnus X-1 system, located in the Cygnus constellation of the Milky Way galaxy. Cygnus X-1 system is a binary system consisting of an invisible compact object and a blue supergiant star, was first detected as an X-ray source in **1964** by **Riccardo Giacconi's team**.

A major turning point came with the launch of the **Uhuru X-ray Observatory** in December **1970.** Launched from the San Marco platform off the coast of Kenya, Uhuru was the world's **first satellite** dedicated entirely **to X-ray astronomy.** Uhuru was also the first NASA spacecraft launched from a foreign location. (The name *Uhuru* means "freedom" in Swahili.)

Orbiting at an altitude of approximately **560 km,** Uhuru operated from 1970 to 1973 and detected **339 cosmic X-ray sources.** These included galaxy clusters, neutron stars, X-ray binaries, and several strong candidates for black holes.

The system emitting strong X-rays from an unseen compact object was first discovered by Riccardo Giacconi and his team. In 1971, the Uhuru satellite observed similar intense X-ray emissions from the same system in the constellation Cygnus, thereby providing strong confirmation of the earlier discovery.

However, X-ray detection alone was not sufficient to prove the existence of a black hole. What was needed was a reliable measurement of the object's mass. Only if the mass exceeded the maximum possible mass of a neutron star—generally taken to be about **3 M⊙,**—could the object be identified as a black hole.

Although Uhuru revealed promising candidates, it could not determine their masses. Its discoveries instead motivated astronomers to carry out detailed follow-up observations, particularly of binary star systems, where orbital motion could be used to measure mass and finally confirm the existence of black holes.

The First Confirmed Black Hole

Motivated by these observations, astronomers **Paul Murdin** and **Louise Webster** in the United Kingdom, and independently **C. T. Bolton** in Canada, carried out detailed optical studies of the visible star associated with Cygnus X-1. They showed that the bright blue supergiant star **HDE 226868** is part of a binary system, orbiting an unseen companion.

Between August and October 1971, Webster and Murdin measured the radial velocity of HDE 226868, while Bolton independently conducted similar spectroscopic observations during the same period.

By carefully analyzing the motion of this star and applying Kepler's laws of planetary motion, they estimated the mass of the invisible object and found it to be far greater than the **maximum possible mass of a neutron star.** This left only one plausible explanation: the unseen companion had to be a **black hole.** Their results, published in 1972, provided the first strong observational evidence for the existence of a stellar-mass black hole.

Figure 1.7: AI-generated portrait of C.T. Bolton (1943-2021) for scientific illustrations

Today, the black hole in the Cygnus X-1 system—now known to have a mass of about **twenty one times that of the Sun**—is widely regarded as the first confirmed black hole, marking a turning point in modern astrophysics.

Figure 1.8: Paul Murdin (Born - 1943), British astronomer

Image Credit: Photographer: Mike Peel/Wikimedia Commons, CC By

Their discovery provided the first convincing observational evidence for an object that had long been considered nearly impossible to detect—a **black hole**. This breakthrough transformed astrophysics and earned the researchers worldwide recognition, professional honours, and lasting scientific acclaim.

Figure 1.9: Artist's impression of Australian astronomer Louise Webster (1941-1990)

Following this landmark discovery, many more black holes have been identified and studied, primarily through advances in X-ray astronomy.

Sadly, **Karl Schwarzschild**, whose solutions to Einstein's field equations first revealed the theoretical possibility of black holes, did not live to witness their observational confirmation. Even **Albert Einstein**, the architect of general relativity who revived earlier ideas related to "dark stars," never saw a black hole conclusively discovered.

In contrast, **Subrahmanyan Chandrasekhar**, once criticised for his revolutionary ideas on stellar collapse, lived to see his theories vindicated. His work demonstrated how sufficiently massive stars could undergo gravitational collapse to form compact objects. For these profound contributions, Chandrasekhar was awarded the **Nobel Prize in Physics in 1983**, marking one of the most significant recognitions in the history of astrophysics.

How Do Astronomers Detect Black Holes

1. X-ray Technology

X-ray technology acts as a distinctive signature, allowing astronomers to detect otherwise invisible black holes. By analysing X-ray emissions, scientists can estimate a black hole's **mass**, **spin**, and **accretion rate.**

The first suspected black hole candidate, Cygnus X-1, was identified as a black hole candidate through X-ray observations. Since then, X-ray astronomy has played a key role in locating and studying such objects.

NASA's **Chandra X-ray Observatory**, launched in 1999, has discovered and examined numerous black holes and continues to operate successfully. More recently, NASA's **Imaging X-ray Polarimetry Explorer (IXPE)**, launched in 2021, has provided new insights into the behaviour of matter and magnetic fields near black holes.

Although X-ray observations are essential for identifying black hole candidates, X-ray emission alone does not conclusively prove the presence of a black hole. Many other astronomical objects can emit X-rays under certain conditions, including neutron stars. Neutron stars and black holes share several similarities, which can make them difficult to distinguish:

- Both are extremely massive.
- Both possess extremely strong gravitational fields due to compactness.
- A neutron star may also form an accretion disk if it is drawing matter from a companion star.
- Both objects can exhibit intense magnetic activity and jet emission.

Isolated neutron stars are extremely small and faint, making them difficult to detect unless they emit radio waves (as pulsars) or X- rays.

Distinguishing Black Holes from Neutron Stars

Astronomers rely on several key differences to tell black holes and neutron stars apart:

1. **Surface Analysis:** Neutron stars have a **solid surface**. As they cool, they emit relatively **weaker X-rays** through thermal radiation. Black holes, by contrast, have **no surface**. Their X-ray emissions originate from the hot accretion disk near the event horizon and are often much more energetic, commonly referred to as **hard X-rays**.
2. **X-ray Spectrum Analysis:** X-rays near a neutron star can interact with its **solid surface**, which slightly changes their **energy and pattern.**

A black hole has no surface, so its X-rays come only from the **hot gas** around it and look different. This difference in X-ray behavior helps astronomers tell neutron stars and black holes apart.

3. **Variability in X-ray Emission:** Black hole systems often show stronger and more rapid variability in X-ray brightness than neutron star systems. This rapid variability provides another important clue to their true nature.
4. **X-ray Bursts and Their Significance:** X-ray bursts are sudden, repeated flashes of X-rays seen in some neutron stars that belong to binary systems. They happen when gas from a nearby companion star builds up on the neutron star's solid surface and suddenly ignites in a powerful nuclear explosion.

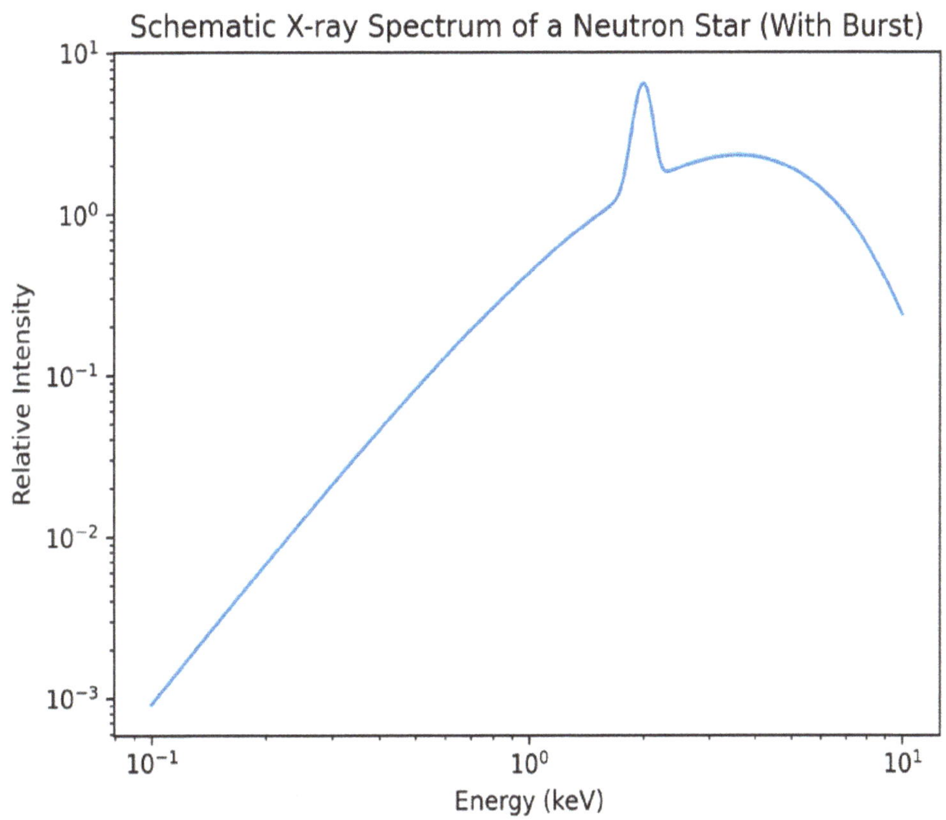

Figure 1.10: Schematic X-ray spectrum of an accreting neutron star showing thermal emission together with a sharp Type I X-ray burst, produced by thermonuclear explosions on the stellar surface.

(These spectra are simplified representations intended for conceptual understanding).

Black holes, however, do not possess a physical surface and cannot support thermonuclear reactions. As a result, true X-ray bursts are never observed from black holes. The absence of X-ray bursts therefore serves as an important diagnostic clue when distinguishing black holes from neutron stars.

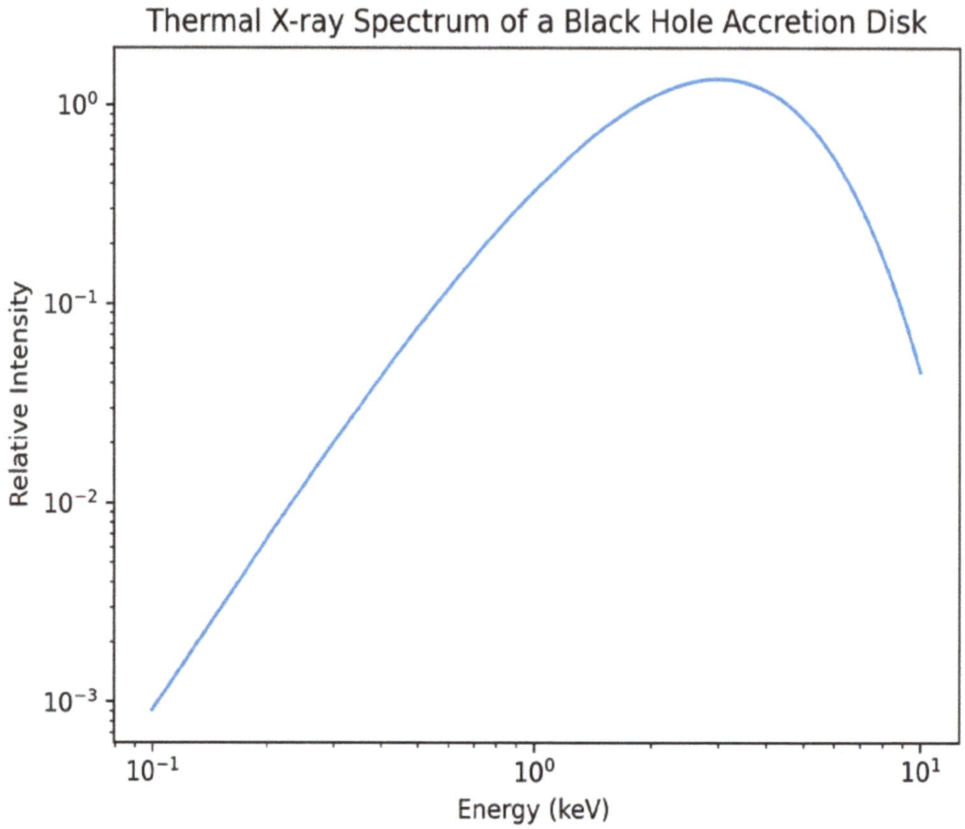

Figure 1.11: Schematic X-ray spectrum of an accreting black hole showing smooth thermal emission from the accretion disk. (These spectra are simplified representations intended for conceptual understanding)

The mere detection of X-rays does not prove the presence of a black hole; however, careful analysis of the full X-ray spectrum allows astronomers to differentiate between black holes and neutron stars.

Early missions such as rocket-launched X-ray detectors and the Uhuru X-ray Observatory detected powerful X-ray sources but could not conclusively confirm black holes.

Estimating Black Hole Mass from X-ray Emission: X-rays originate from hot gas falling toward a black hole. The way these X-rays change in brightness and energy tells astronomers how strong the black hole's gravity is. From this, they can estimate its mass.

Measuring Black Hole Spin Using X-rays: The spin of a black hole affects how close hot gas can orbit before falling in. X-ray observations reveal how the hot gas behaves near the black hole's event horizon, especially through changes in X-ray energy and shape of spectral lines. By studying these features, astronomers can estimate how fast the black hole is spinning.

Limitations of X-ray Techniques in Identifying Black Holes

Despite being one of the most powerful tools for detecting and studying black holes—and playing a key role in their earliest discoveries—X-ray technology has several important limitations.

1. **Detection of Rogue Black Holes:** X-ray techniques are **incapable** of **detecting wandering (isolated) black holes** or **dormant black holes**, as these objects do not actively accrete matter and therefore lack strong X-ray emission.

2. **Mass Measurement:** X-ray observations alone are not sufficient to determine the true nature or mass of a black hole. While X-rays reveal the presence of hot gas falling toward a compact object, they do not directly measure its mass. To estimate mass, astronomers must combine X-ray data with several other techniques, such as studying the motion of a companion star, orbital dynamics, Doppler shifts, and, in some cases, gravitational-wave observations. Even when multiple methods are used together, the mass of a black hole cannot be measured with perfect precision. Instead, astronomers usually determine a range of possible masses, rather than a single exact value. This makes mass measurement one of the most difficult and uncertain tasks in black hole research.

3. **Dependence on Active Accretion:** X- ray detection requires hot, infalling matter. If accretion is weak or intermittent, X-ray emission may fall below detection limits, causing even known black hole systems to appear inactive or disappear temporarily from view.

4. **Obscuration by Gas and Dust:** Although X-rays penetrate dust better than visible light, dense gas clouds—especially near galactic centers—can absorb or scatter X-rays, reducing signal strength and complicating interpretation.

5. **Limited Spatial Resolution:** X-ray telescopes cannot directly image the event horizon or resolve fine structures close to a black hole. Most observed X-ray signals come from surrounding hot gas, not the black hole itself.

6. **Variability and Transient Behavior:** X-ray sources often vary rapidly in brightness. While variability provides clues, it can also make long-term interpretation difficult, as short observations may capture unusual or unrepresentative states.

7. **Contamination from Other X-ray Sources:** Regions rich in stars, supernova remnants, or hot gas can produce overlapping X-ray signals, making it challenging to isolate emission from a single black hole candidate.

8. **Distance Uncertainty:** X-ray observations alone cannot accurately determine how far away a source is. Distance errors directly affect estimates of luminosity and physical size, leading to uncertainties in interpretation.

9. **Instrumental Sensitivity Limits:** Even the most advanced X-ray observatories have sensitivity thresholds. Very faint or distant X-ray sources may remain undetectable, especially in other galaxies.

10. **Model Dependence:** Interpreting X-ray data relies heavily on theoretical models of accretion disks and plasma physics. Different models can sometimes explain the same data, leading to ambiguous conclusions.

Therefore, astronomers rely on complementary techniques such as astrometry, which measures the precise motion of stars influenced by an unseen massive object, and observations of visible light from surrounding stars and gas. In special cases, gravitational wave detections from merging black holes provide additional confirmation. When combined, these methods offer a more complete and reliable way to identify and study black holes across different environments.

Other Methods to Detect Black Holes

1. Gravitational Lensing

Objects with enormous mass, such as black holes, can **bend and distort light** from distant background stars. This effect, predicted by Einstein's theory of general relativity, is known as **gravitational lensing**.

Gravitational lensing is especially valuable for detecting **isolated** or **wandering black holes**, which are otherwise invisible. When such a black hole passes in front of a distant star, it briefly magnifies or shifts the star's apparent position, revealing the black hole's presence.

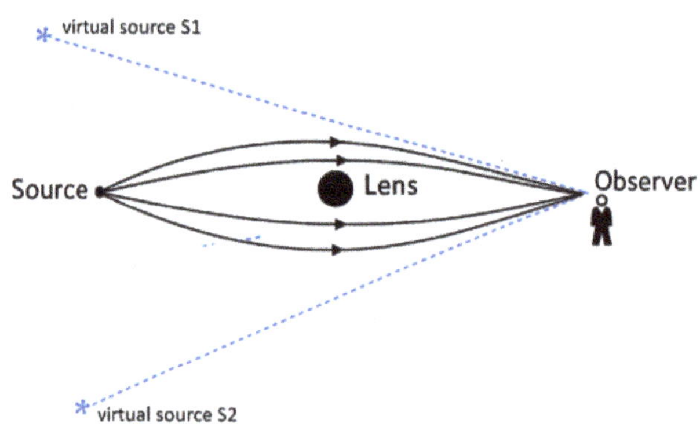

Figure 1.12: Gravitational lensing illustration

2. Astrometry and the Discovery of Black Holes

Astrometry is the branch of astronomy devoted to the precise measurement of the positions, motions, and orbits of celestial objects. In recent years, it has become a powerful method for detecting black holes that emit little or no radiation.

Although a black hole itself cannot be seen, its gravity can betray its existence. When a black hole is part of a binary system, its gravitational pull causes the visible companion star to exhibit a subtle **"wobble"** in its motion. By carefully tracking this movement over time and applying **Kepler's laws of motion**, astronomers can determine the mass of the unseen companion.

If the inferred mass exceeds the maximum possible mass of a **neutron star (about 2–3 M⊙), t**he most plausible explanation is that the unseen object is a **black hole**.

Astrometry is particularly important for identifying **dormant black holes**, which are not actively accreting matter and therefore remain invisible to X-ray telescopes. Instead, their presence is revealed through their gravitational influence on nearby stars.

Recent advances—most notably the **Gaia mission** of the European Space Agency—have revolutionised astrometry. Gaia tracks the motions of more than a billion stars with extraordinary precision, enabling astronomers to detect minute stellar wobbles caused by unseen black hole companions. This has already led to the identification of several black hole candidates within the Milky Way galaxy.

Astrometry thus represents a transformative approach in black hole astronomy, complementing X-ray observations, gravitational-wave detections, and radio studies to build a more complete picture of the black hole population.

3. Bright Jets of Material

In some systems, matter falling toward a black hole is not entirely swallowed. Instead, a portion of it is expelled in the form of **relativistic jets**—narrow beams of high-energy particles launched at nearly the speed of light.

Figure 1.13: AI-generated illustration of jets of black hole for conceptual purposes

These powerful jets can be detected across vast cosmic distances and serve as strong evidence of **active black hole accretion**. While some neutron stars can produce jets, they are generally far less powerful than those associated with black holes.

4. Gravitational Waves: The Discovery of Merging Black Holes

When massive objects undergo rapid acceleration, they produce gravitational waves—ripples in the fabric of spacetime predicted by Einstein's theory of general relativity. These waves are especially strong during catastrophic events such as the merger of two black holes or two neutron stars.

When two massive objects orbit each other, they emit gravitational waves long before they finally collide. As the objects spiral closer together, these waves gradually increase in strength and frequency, producing a characteristic signal known as the **inspiral.** From this early part of the signal, astronomers can measure the masses of the two objects with high accuracy.

Figure 1.14: AI-generated illustration of merging of black holes and gravitational waves for conceptual purposes

If the measured masses are greater than the maximum possible mass of a neutron star, the objects must be black holes. In this way, black holes can be identified even before the merger occurs, purely from the gravitational-wave pattern they produce.

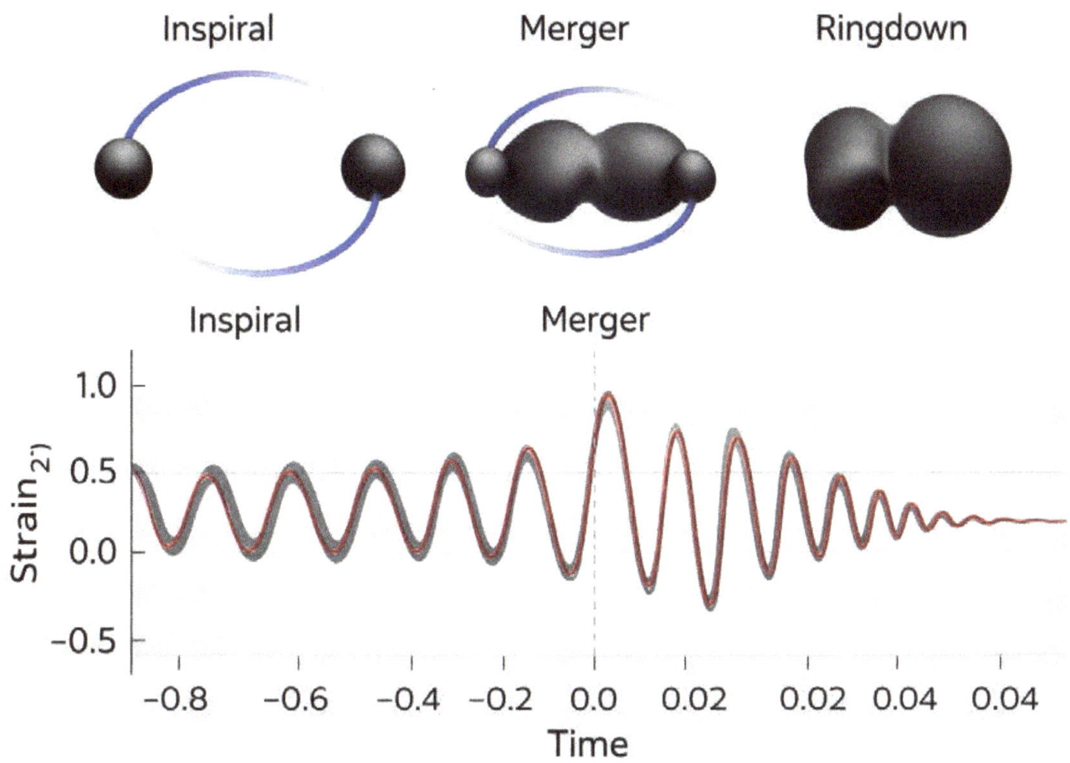

Figure 1.15: Schematic representation of a binary black hole merger, illustrating the inspiral phase, merger, and ringdown, accompanied by the emission of gravitational waves

Image Credit: Original illustration for explanatory purposes.

After the merger, the newly formed object briefly vibrates in a phase called the **ringdown,** creating another distinctive gravitational-wave signal. This final signal confirms that the merged object is a black hole. Together, the inspiral and ringdown phases allow astronomers to detect and identify black holes using gravitational waves alone, even when no light is emitted.

Although gravitational waves are extremely weak by the time they reach Earth, they can be detected by highly sensitive instruments such as **LIGO** (Laser Interferometer Gravitational-Wave Observatory) and **Virgo.** Since 2016, these observatories have detected many black hole mergers, opening an entirely new way to study the universe. Gravitational-wave astronomy allows scientists to observe cosmic events that are completely invisible to traditional telescopes.

The LIGO–Virgo–KAGRA (LVK) Gravitational-Wave Network

The LIGO–Virgo–KAGRA (LVK) network is an international collaboration of gravitational-wave detectors designed to observe mergers of the universe's most compact and extreme objects. The network consists of observatories located in the United States, Europe, and Japan, working together to detect tiny ripples in spacetime produced when massive objects collide.

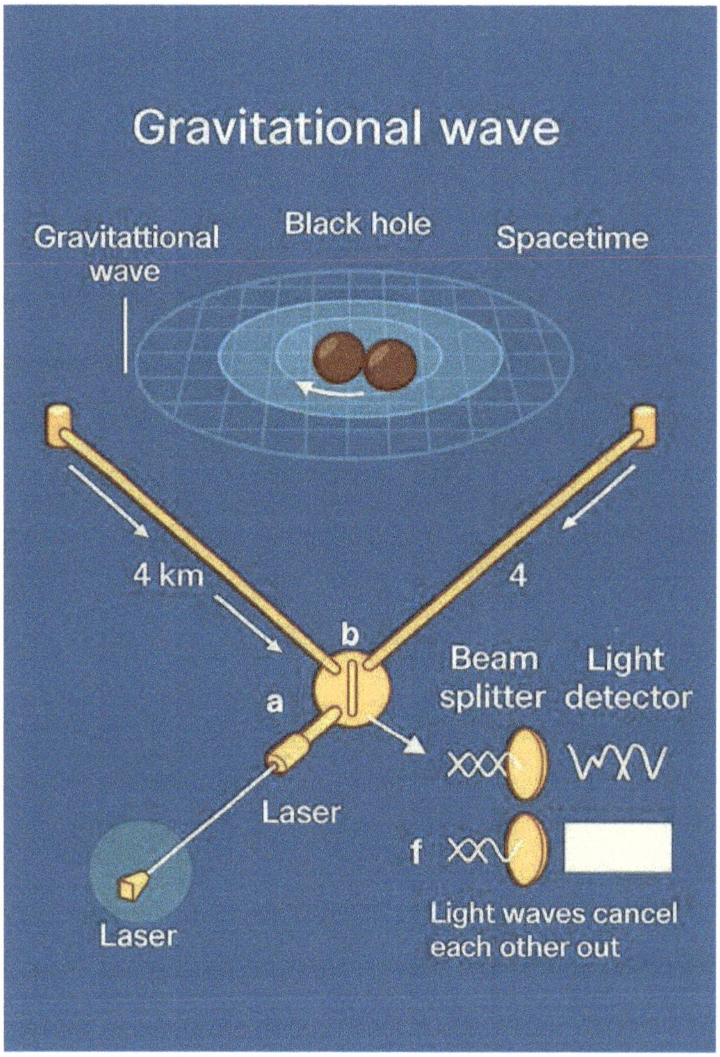

Figure 1.16: AI-generated illustration of Ligo Observatory to detect gravitational waves for explanatory purposes

The first gravitational-wave signal was detected on **14 September 2015, marking** the beginning of gravitational-wave astronomy. Since then, through successive observing runs with steadily improving sensitivity, the LVK network has detected nearly **300 gravitational-wave events.**

Among these detections

1. Around **200 events** are confirmed **binary black hole mergers**, in which two black holes spiral together and merge.
2. **Two events** confirmed **black hole–neutron star mergers**, representing rare mixed systems.
3. **Two events** confirmed **binary neutron star** mergers, including the landmark event that also produced visible light.

The remaining events are classified as candidate detections, whose exact nature is still under detailed analysis and has not yet been confirmed with the highest confidence.

These discoveries have revealed that black hole mergers are far more common than once expected and occur across vast cosmic distances and timescales.

Significance: The LVK network acts as a powerful tool for uncovering merging black holes and neutron stars, transforming invisible cosmic catastrophes into observable signals and greatly expanding our knowledge of the universe's most extreme objects and the nature of gravity itself.

The **2017 Nobel Prize** in **Physics** was awarded to **Rainer Weiss**, **Barry C. Barish,** and **Kip S. Thorne** for decisive contributions to the LIGO detector and the observation of gravitational waves.

Conclusion: Over the past century, astronomers have transformed the once-daunting challenge of detecting invisible black holes into a rigorous and reliable scientific endeavour. Through a combination of observational techniques—including X-ray astronomy, gravitational lensing, astrometric measurements, relativistic jets, and gravitational-wave detections—scientists can now identify black holes across a wide range of masses and cosmic environments. In many cases, two or more of these methods are used together to confirm a black hole's existence with high confidence.

These approaches not only reveal the locations of black holes but also allow researchers to determine their **masses**, **spins**, and the strength of their **gravitational influence** on surrounding matter and spacetime itself. What began as a purely theoretical prediction has evolved into one of the most powerful and sophisticated fields in modern astrophysics, offering profound insights into the life cycles of stars, the structure of galaxies, and the fundamental nature of the universe.

Chapter 2

The Concept of Black Holes and Their Formation

Introduction

Black holes are among the most mysterious and fascinating objects in the universe. They represent the ultimate triumph of gravity—regions where spacetime is warped so extremely that nothing, not even light, can escape. Despite their dramatic nature, black holes are not sudden or exotic anomalies; they form naturally as a consequence of stellar evolution and the fundamental laws governing matter, energy and spacetime.

This chapter explores how the concept of black holes evolved from a theoretical idea into a central pillar of modern astrophysics. It also explains how these extraordinary objects form through the gravitational collapse of matter, revealing the physical processes that transform massive stars into some of the most extreme entities known to science.

What is a Black Hole

A black hole is an astronomical object so compact and massive that its gravitational pull prevents anything from escaping—even light. It forms when a large amount of matter is compressed into an extraordinarily small region, causing gravity to become extremely strong.

In such a region, spacetime itself is intensely curved. Matter and radiation that cross a certain boundary become permanently trapped. Thus, a black hole can be defined as a region of spacetime where gravity has become so powerful that escape is impossible.

Why is it Called a "Black Hole"

A black hole emits no light of its own. Since light cannot escape its gravitational pull, the object appears completely dark against the background of space.

In literature, the **colour black symbolizes mystery, depth, and power**–ideas that harmonize perfectly with the scientific nature of this cosmic phenomenon.

The word "**hole**" suggests a region without any solid or liquid surface, where matter and energy seem to vanish beyond view.

Thus, the term **"black hole"** beautifully unites the language of **science** and the symbolism of **literature**—a name that conveys both the mystery of its unseen depths and the immense power that defines it. The expression was first popularized in 1967 by American theoretical physicist **John Archibald Wheeler** of Princeton University, who used it to describe these extraordinary regions of space where gravity reigns supreme.

The concept once described as a "**dark star"** by Pierre-Simon Laplace in the 18th century later came to be known as a "**black hole."**

A black hole is generally described by two fundamental components:

1. **The singularity (core)**
2. **The event horizon**

1. Singularity (core)

The term *singularity* comes from the Latin *singularis*, meaning "single" or "unique," reflecting its nature as a point unlike any other in the universe.

A black hole is not empty. At its center lies the singularity—an extremely compact region where an enormous amount of matter is compressed into an infinitesimally small volume. According to classical general relativity, the singularity has infinite density and zero volume. It has no solid surface, no recognizable structure, and cannot be observed directly.

Under such extreme conditions, familiar forms of matter—atoms, nuclei, protons, and electrons—cease to exist in any known state. All physical structure collapses, and even the usual concepts of space and time lose their meaning.

Although the internal nature of the singularity remains unknown, its total mass continues to influence the surrounding universe through gravity, shaping the curvature of spacetime outside the black hole. The singularity represents one of the deepest unsolved problems in physics, where Einstein's equations predict infinities and ultimately break down. A complete explanation likely requires a quantum theory of gravity, which has yet to be formulated.

2. Event Horizon

The event horizon of a black hole is an invisible boundary that surrounds and encloses all the matter and energy that make up the black hole. It is not a physical surface like that of the Earth or even the Sun. Once any object, particle, or even light crosses this boundary, it can never return to the outside universe. For distant observers, the event horizon marks the ultimate limit beyond which information cannot be recovered.

Escaping from inside a black hole requires infinite energy, but falling into one requires no minimum kinetic energy. Matter with no angular momentum will naturally fall inward under gravity, while rotating matter must first lose angular momentum before crossing the event horizon.

In literature, an event horizon is often described metaphorically as a point of no return — a boundary separating the known from the unknown. This poetic image closely mirrors its physical meaning in astrophysics.

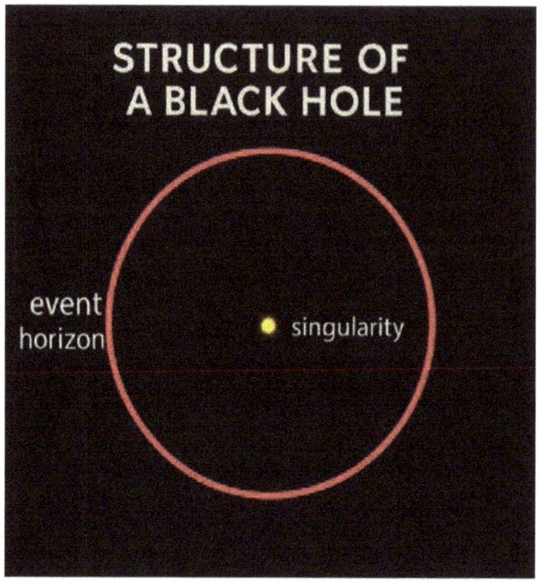

Figure 2.1: AI-generated conceptual illustration based on standard model in general relativity

Structure of a Black Hole

A black hole consists of a singularity enclosed by an event horizon. While matter may exist inside this boundary, we currently have no way to observe or directly study it. What happens beyond the event horizon remains hidden from the outside universe, since no information or signal can escape once it crosses this boundary.

Black Hole Regions and the Schwarzschild Radius

The size of a black hole is defined by the radius of its event horizon, known as the **Schwarzschild radius** (for a non-rotating black hole):

$$R_s = 2GM/c^2$$

Where:

Rs = Schwarzschild radius

G = gravitational constant

M = mass of the black hole

c = speed of light

Key Properties

- The Schwarzschild radius is **directly proportional to mass**.
- A useful approximation is **3 km per solar mass**.

Examples

A **10 M⊙ black hole** has an event horizon radius of about **30 km**.

A **supermassive black hole** with **1 million M⊙** has an event horizon radius of about **3 million km**, roughly twice the Sun's diameter.

If the **Sun** (radius ≈ 700,000 km) collapsed into a black hole, its event horizon would have a radius of only about **3 km**.

If the **Earth** (radius ≈ 6371 km) were to collapse into a black hole, its Schwarzschild radius would be only about **9 millimetres**.

This astonishing fact shows how unimaginably concentrated matter becomes inside a black hole, where an entire planet's mass is compressed into an extraordinarily tiny region of space. Though it may seem unbelievable, this is a direct and well-established consequence of Einstein's theory of gravity.

Information Loss Paradox

According to classical physics, any object or information falling into a black hole appears permanently lost to outside observers. This idea leads to the **black hole information loss paradox**, which challenges fundamental principles of quantum mechanics. Resolving this paradox remains one of the most important open questions in theoretical physics.

Regions Around a Black Hole

1. Accretion Disk of Black Hole

A black hole can be surrounded by rings of gas and dust called an accretion disk. An accretion disk is a flattened, rotating disk of matter that forms around a black hole and lies outside its event horizon. As material is pulled toward the black hole by its immense gravity, it does not fall straight in. Instead, due to the conservation of angular momentum, the material spirals inward, forming the disk.

Friction and gravitational forces within the disk cause the infalling material to accelerate to extremely high speeds. This rapid motion generates intense heat, raising the temperature of the gas to millions of degrees. As a result, the accretion disk becomes a powerful source of radiation.

Corona

Above the disk lies a compact region of extremely hot plasma, known as the **corona,** which is energized by the disk and acts as a strong source of X-ray emission.

Radiation

Around stellar-mass black holes, this radiation is mainly emitted as X-rays, while around supermassive black holes it can appear as ultraviolet and visible light. These high-energy emissions allow astronomers to detect black holes even though the black holes themselves emit no light.

From Orbit to Plunge

As matter moves closer to the black hole, it gradually loses angular momentum and spirals inward. Near the inner edge of the disk, stable orbits cease to exist, and matter plunges rapidly toward the event horizon. No special energy is required to cross this boundary; gravity alone drives the infall.

Black Hole Jets

In many systems, part of the infalling material is redirected into powerful jets that shoot out along the black hole's rotational axis at nearly the speed of light. These jets, produced by strong magnetic fields interacting with the disk and the black hole's rotation, can extend across enormous distances and emit radiation from radio waves to gamma rays.

Because of their intense radiation, variability, and jet activity, accretion disks are one of the most powerful tools for detecting and studying black holes. By analyzing their light, astronomers can estimate a black hole's mass, spin, and rate of growth, making accretion disks key to understanding black holes and their influence on the universe.

Figure 2.2: AI-generated illustration of a black hole and its surrounding accretion disk for conceptual purposes

2. Innermost Stable Circular Orbit (ISCO)

The innermost stable circular orbit (ISCO) marks the closest distance at which matter can maintain stable orbit around a black hole safely. Inside this boundary, stable orbits disappear, and matter rapidly falls toward the black hole.

For a non-rotating (Schwarzschild) black hole:

$r_{ISCO} = 6GM/c^2$

This corresponds to **three times the Schwarzschild radius (3 r_s),** or **$6GM/c^2$.**

3. Photon Sphere

The **photon sphere** is a region where gravity is strong enough that light itself can orbit the black hole.

For a non-rotating black hole

$r_{ph} = 3GM/c^2$

For a Schwarzschild black hole, **the photon sphere lies at 1.5 times the Schwarzschild radius.**

Although photons can briefly follow circular paths here, these orbits are **unstable**. Even a tiny disturbance causes the photon to either fall into the black hole or escape outward.

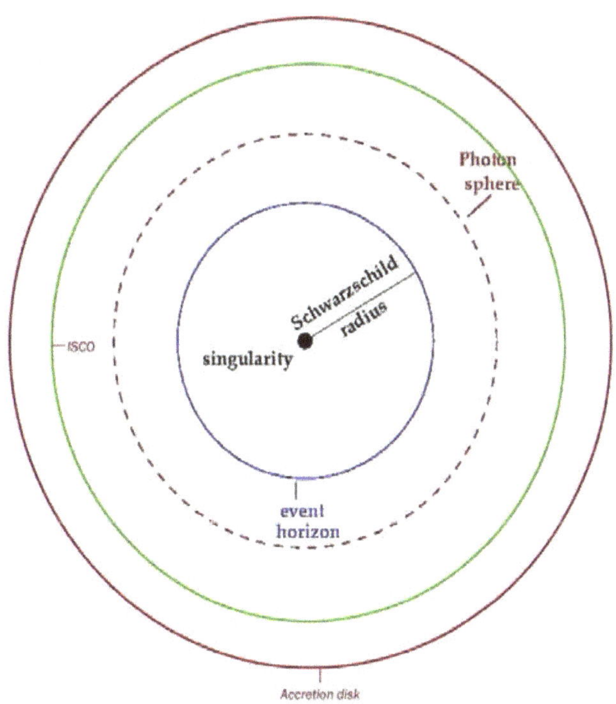

Figure 2.3: Artist's impression illustrating the main regions surrounding a black hole

Relationship Between Key Radii

For a non-rotating black hole:

$r_s = 2GM/c^2$

Thus $r_s < r_{ph} < r_{ISCO}$

The **event horizon** lies closest to the singularity. The **photon sphere** lies outside the event horizon. The **ISCO** lies farther out and defines the inner edge of the accretion disk.

4. Ergosphere (Rotating Black Holes)

The term ergosphere is derived from the Ancient Greek word "*ergon,*" meaning "*work*:" or "*energy.*" The name reflects a remarkable property of rotating black holes. Within this region, it is theoretically possible to extract energy from the black hole's rotation.

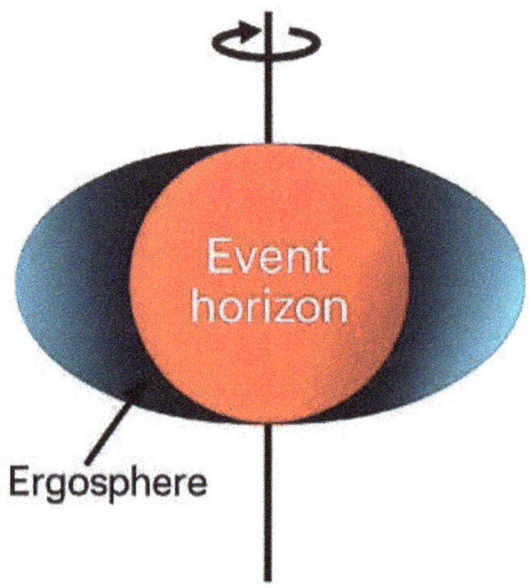

Figure 2.4: AI-generated illustration of a rotating black hole showing the event horizon and surrounding ergosphere for conceptual purposes

A rotating black hole is known as a **Kerr black hole,** named after the New Zealand mathematician **Roy Kerr,** who discovered the exact solution describing a rotating black hole in 1963.

Unlike a non-rotating black hole, a Kerr black hole possesses a unique region called the **ergosphere**, which exists only around rotating black holes. This region lies just outside the event horizon and is bounded externally by the **static limit**.

Within the ergosphere, objects cannot remain stationary relative to distant observers. Even light can not remain stationary in this region ; it is compelled to move in the direction of the black hole's rotation.

This unusual behavior arises from frame dragging, an effect predicted by Einstein's general theory of relativity. A rotating mass drags the surrounding spacetime along with it. Near a rapidly spinning black hole , this effect becomes extremely strong, forcing all matter and radiation within the ergosphere to co-rotate with the black hole.

A Kerr black hole possesses two horizons: an outer event horizon, and an inner horizon, known as the Cauchy horizon.

The ergosphere lies outside the outer event horizon, forming a region where spacetime rotation becomes so intense that no object can remain at rest with respect to distant observers.

Accretion Disk vs. Ergosphere: Distinct Regions

Although often confused, the ergosphere and the accretion disk are fundamentally different regions: The ergosphere is a region of warped spacetime created by rotation; it contains no matter by definition.

The accretion disk consists of real matter—gas and dust—orbiting the black hole outside the event horizon.

Typically, there is a vacuum gap between the outer boundary of the ergosphere and the inner edge of the accretion disk. However, if a black hole spins extremely rapidly (approaching the theoretical maximum), the inner edge of the accretion disk can move inward and partially overlap the outer boundary of the ergosphere. In such cases, disk material directly experiences intense frame dragging, allowing energy-extraction mechanisms to operate efficiently.

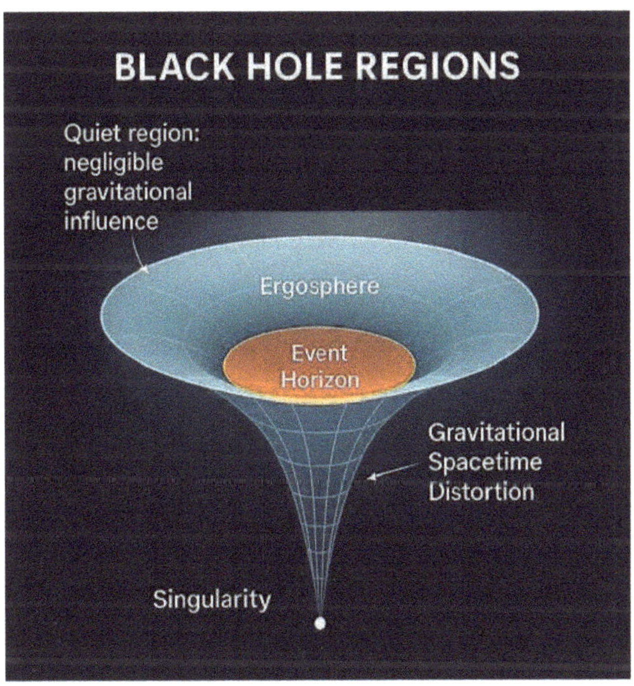

Figure 2.5: AI-generated illustration of Kerr rotating black hole for explanatory purposes

The event horizon marks the ultimate boundary of the black hole, tightly wrapped by the ergosphere, which hugs it most closely at the poles. The accretion disk orbits farther out, sometimes brushing the ergosphere only when the black hole spins at its most extreme rate.

Prograde Orbit: The black hole's spin is in the same direction as the orbiting matter.

Retrograde Orbit: The black hole's spin is opposite to the direction of the orbiting matter. This distinction strongly affects orbital stability and energy extraction.

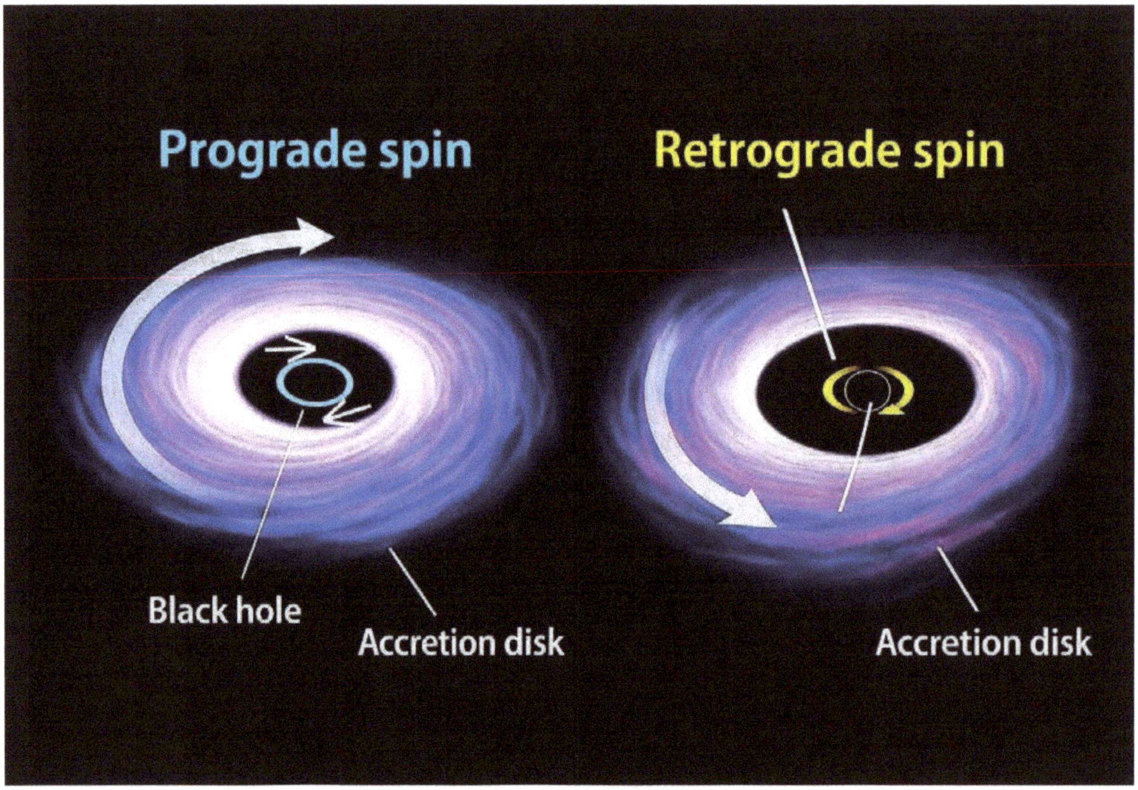

Figure 2.6: AI-generated illustration of prograde and retrograde spin for conceptual purposes

ISCO and Photon Ring in Relation to the Ergosphere

Retrograde Orbit: Both the ISCO (innermost stable circular orbit) and the photon ring lie outside the ergosphere.

Prograde Orbit: For moderate spin, both the ISCO and photon ring remain outside the ergosphere.

For rapid spin, parts of the ISCO and photon sphere may lie inside the ergosphere, allowing orbiting matter and light to experience strong frame dragging.

Formation of Black Holes

1. Stellar Evolution

Life Cycle of a Massive Star

Stars like the Sun spend most of their lives fusing hydrogen into helium in their cores through nuclear fusion. This process generates outward pressure that balances the inward pull of gravity, maintaining the star's stability.

In **massive stars**, nuclear fusion continues through a series of heavier elements. After hydrogen is exhausted, the core successively fuses:

- Helium,
- Carbon,
- Neon,
- Oxygen,
- Silicon.

During the silicon-burning phase, silicon nuclei undergo a series of nuclear reactions that gradually build heavier elements such as sulfur, argon, and calcium. In the final stage of this process, these reactions produce **iron,** which accumulates in the core. Since fusion of iron does not release energy, nuclear burning can no longer support the star.

With no outward pressure remaining to counter gravity, the core collapses catastrophically.

1. If the core mass is less than about **1.44 M⊙,** it cannot collapse further and forms a **white dwarf.** For that initial mass of the star lies in the range of **0.8 to 8 M⊙**
2. If the collapsing core has a mass between about **1.44** and **2–3 M⊙,** it forms a **neutron star.** These remnants originate from stars whose initial masses are roughly **8–20 M⊙.**
3. However when the core of a dying star becomes heavier than about **1.4 times the mass of the Sun, it can no longer remain a white dwarf**. It collapses under gravity, and if the mass is not too large, it forms a neutron star. These neutron stars typically have masses around **1.1 to 2 times that of the Sun.**
4. Even if they have the same mass, a white dwarf is spread over a planet-sized volume, while a neutron star compresses that mass into a city-sized sphere.
5. If the remaining core exceeds **about 3 M⊙,** gravity dominates all known forces and the core collapses into a black hole. The original star must typically have a mass of

20–30 times M⊙ while the outer layers of the star are expelled in a powerful supernova explosion. Only the most massive stars end their lives as black holes.

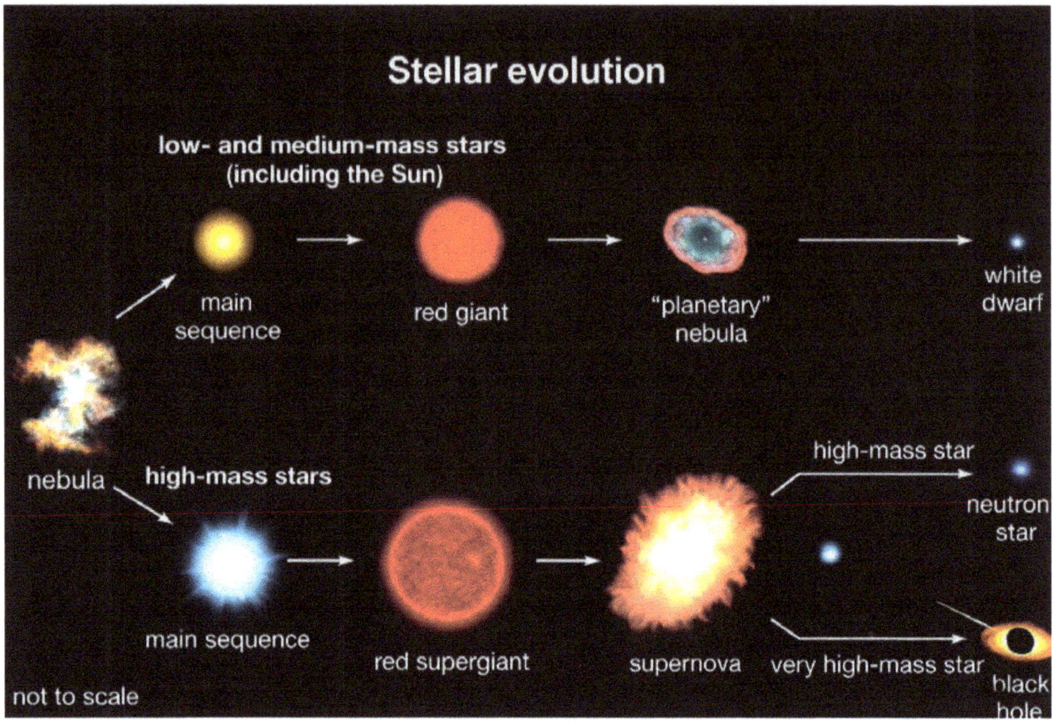

Figure 2.7: Conceptual AI-generated illustration based on the standard model of stellar evolution.

Most black holes currently observed belong to this category and are known as stellar-mass black holes. Thus, we conclude that all stars do not have the potential to become black holes.

2. Primordial Black Holes

Primordial black holes are hypothetical black holes that may have formed in the early universe from extremely large **density fluctuations shortly after the Big Bang.**

Primordial black holes are hypothetical objects that may span an enormous range of masses—from microscopic to supermassive. Detecting or constraining their existence could provide important insights into early-universe physics and the nature of dark matter.

3. Direct-Collapse Black Holes

Direct-collapse black holes are thought to form when enormous clouds of gas in the early universe collapse directly under their own gravity, bypassing the usual stage of star formation.

They are not purely hypothetical; many astronomers consider them the most likely seeds of the supermassive black holes observed at the centers of galaxies.

4. Microscopic Black Holes

4.1. Microscopic Black Holes from Particle Collisions (Artificial)

Microscopic black holes could theoretically be produced by extremely high-energy particle collisions, including interactions involving cosmic rays or neutrinos

Some theoretical models suggest that extremely high-energy particle collisions could create microscopic black holes. However, the required energies are far beyond current technological capabilities.

The minimum energy required is estimated at around **10^{16} TeV.** The Large Hadron Collider (LHC) operates at: **14 TeV for proton–proton collisions ~1,150 TeV for lead–ion collisions.**

These energies are insufficient to create black holes in any realistic scenario.

4.2. Microscopic Black Holes from Cosmic Rays (Natural)

Certain advanced theories propose that ultra-high-energy cosmic rays or neutrinos could, under special conditions, create microscopic black holes in Earth's atmosphere—particularly if extra spatial dimensions exist and gravity becomes stronger at the TeV scale. Some of these particles carry energies exceeding 10^{20} electron volts, equivalent to tens of millions of tera–electron volt.

If such black holes were formed, they would be extremely short-lived, evaporating almost instantly through Hawking radiation. To date, neither microscopic black holes nor Hawking radiation have been experimentally detected, and the underlying assumptions of these theories remain unverified.

Conclusion

Black holes are defined by only two fundamental components: the singularity and the event horizon. Surrounding this boundary are several important regions—such as the photon sphere, ergosphere, and innermost stable circular orbit (ISCO)—along with structures like the accretion disk and corona, which together shape their observable properties. Their formation arises through multiple pathways, most prominently the gravitational collapse of massive stars, but also through more exotic possibilities such as primordial, direct-collapse and microscopic black holes proposed by theory. Understanding both the internal structure and the diverse formation mechanisms of black holes provides the foundation for exploring their astrophysical roles, observational detection, and profound implications for gravity and spacetime.

Chapter 3
Types and Classification of Black Holes

Introduction

Although black holes are governed by remarkably simple physical principles, they display an extraordinary diversity in their properties and origins. In general relativity, a black hole is fully characterized by only a few fundamental parameters—**mass, electric charge, and spin (angular momentum)**. Yet, these few quantities give rise to objects that range from stellar remnants only a few kilometers across to supermassive giants millions or billions of times more massive than the Sun, residing at the center of galaxies.

Classifying black holes provides a systematic framework for understanding their physical nature, formation mechanisms, and astrophysical roles. Broadly, black holes are classified in two principal ways:

1. **By Physical Parameters:** Based on the fundamental quantities that define their spacetime geometry—**mass, electric charge, and spin.**
2. **By Mass Scale**: Based on their mass range and origin, black holes are divided into **stellar-mass**, **intermediate-mass,** and **supermassive categories**. The first approach reflects the theoretical structure of black holes as predicted by general relativity, while the second highlights their observational diversity across the universe.

Together, these classifications form the foundation for understanding one of nature's most extreme and fascinating phenomena.

The Theoretical Structure of Black Holes: The No-Hair Theorem

The term "No-Hair Theorem" was popularized by **John Archibald Wheeler** in the late 1960s, but the theorem itself was developed through the work of several physicists, including Werner Israel, Brandon Carter, and David Robinson. This phrase means that black holes have no "hair" and can be completely described by only three externally observable physical properties:

- **Mass**
- **Electric charge**
- **Spin (angular momentum)**

All other information about the matter that formed the black hole—its composition, structure - or history—is lost beyond the event horizon. Consequently, black holes are classified primarily according to these three physical characteristics.

1. Simple Black Holes: Schwarzschild Black Holes

The simplest theoretical black holes are those that possess **mass only**, with **no electric charge** and **no rotation**. These are known as **Schwarzschild black holes.**

They are:

- Perfectly spherical
- Static (non-rotating)
- Described by exact spherical symmetry
- Characterized by a single event horizon and a point-like singularity at the center

They exist in empty spacetime without surrounding matter.

However, Schwarzschild black holes are **purely theoretical constructs**. No such objects have been observed, and they are unlikely to exist in nature. In realistic astrophysical environments, black holes form through the gravitational collapse of massive stars—a process that conserves angular momentum. As a result, nearly all real black holes are expected to rotate.

The Schwarzschild solution is a vacuum solution invaluable for understanding basic black hole physics; it represents an idealized limit rather than a physically realized object.

2. Kerr Black Holes: Rotating Black Holes

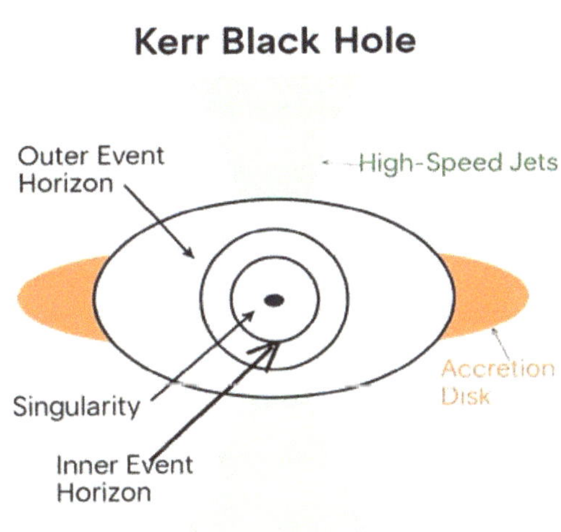

Figure 3.1: AI-generated illustration of Kerr black hole showing its various regions for explanatory purposes.

Kerr black holes are rotating black holes that possess **mass and angular momentum** but no electric charge.

Kerr black holes are described by a solution to **Einstein's Field Equations**, discovered by **Roy Kerr in 1963**.

Figure 3.2: Roy Kerr (1934-2026), New Zealand astronomer and mathematician

Image credit: Bengt Nyman, Wikimedia Commons, CC BY-SA 4. 0

Kerr black holes are believed to be the most common type of black hole in the universe, formed from the collapse of rotating massive stars or through the merger of black holes.

Key Characteristics

Unlike Schwarzschild black holes, Kerr black holes exhibit several unique features:

Ring Singularity: Instead of a point singularity, Kerr black holes contain a **ring-shaped singularity**, arising from their rotation.

Two Horizons: Kerr black holes possess an **outer event horizon**, beyond which nothing can escape, and an **inner (Cauchy) horizon**, a mathematically predicted surface that is believed to be classically unstable.

Oblate Event Horizon: Rotation causes the event horizon to become **oblate**—flattened at the poles and extended at the equator.

Frame Dragging: The rotation of a Kerr black hole drags spacetime around it, forcing nearby matter and radiation to move in the direction of rotation.

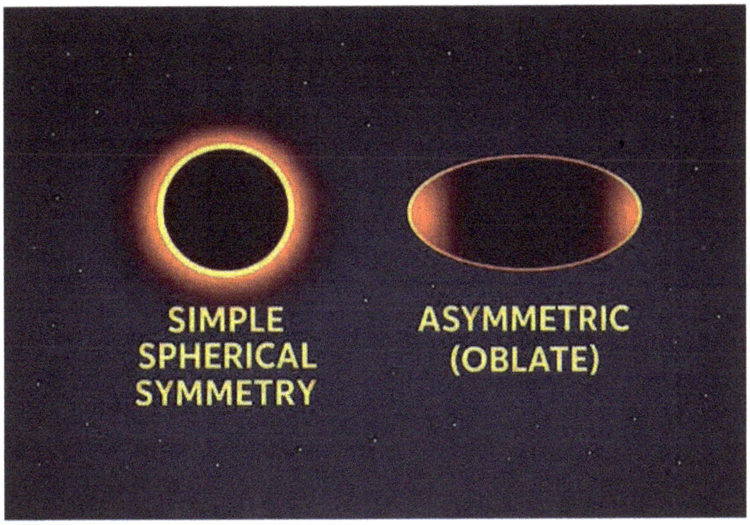

Figure 3.3: AI-generated illustration for conceptual purposes

Formation and Spin Evolution

Black holes acquire and modify their spin through several astrophysical processes:

1. **Stellar Collapse:** When a rotating massive star collapses, angular momentum is conserved. As the star's radius shrinks dramatically, its rotation rate increases, often producing a rapidly spinning black hole.
2. **Accretion:** Matter falling into a black hole can either increase or decrease its spin, depending on the direction of the infalling material relative to the black hole's rotation.
3. **Black Hole Mergers:** When two black holes merge, the resulting black hole is initially distorted and highly dynamic. It then emits gravitational waves during a relaxation phase known as **ringdown**, eventually settling into a stable Kerr black hole. The final spin depends on the masses, spins, and alignment of the original black holes.

Event Horizon Radius and Spin Parameter

For a non-rotating (Schwarzschild) black hole, the event horizon radius is

$$\mathbf{R_s = 2GM/c^2}$$

For a rotating (Kerr) black hole, the radius of the **outer event horizon** is:

$$\mathbf{R_h = GM/c^2 \, [\, 1 + (\, 1 - a^{*2}\,)^{1/2}]}$$

Where:

$\mathbf{R_h}$ = radius of event horizon of a rotating black hole

G = gravitational constant

M = mass of the black hole

c = speed of light

a* = dimensionless spin parameter

If **a* = 0,** for non rotating black hole

$\mathbf{R_h = 2GM/c^2}$

If **a* = 1** (maximum spin parameter)

$\mathbf{R_h = GM/c^2}$

The size of a black hole depends not only on its mass but also on its spin. For the same mass, a rotating (Kerr) black hole always has a smaller event-horizon radius than a non-rotating (Schwarzschild) black hole. In other words, rotation makes a black hole more compact, while slowing the spin causes the event horizon to expand.

The spin parameter is defined as:

$\mathbf{a^* = Jc/GM^2}$

Where **J** is the angular momentum.

The spin parameter ranges from **−1 to +1,** where the sign indicates the direction of rotation, but real black holes cannot reach this limit due to energy losses and physical constraints.

3. Charged Black Holes

In principle, a black hole may also possess **electric charge**, leading to solutions that include mass, charge, and spin. Such black holes are important in theoretical physics, though they are unlikely to exist in nature with significant charge.

Types of Charged Black Holes

1. Reissner–Nordström Black Hole

- Charged
- Non-rotating
- Characterized by mass and electric charge

2. Kerr–Newman Black Hole

- Charged and rotating

- Characterized by mass, charge, and spin

These solutions extend the Schwarzschild and Kerr models and play a role in black hole thermodynamics and quantum gravity research.

Practical Considerations

Astrophysical black holes are expected to be **effectively electrically neutral**. Electric forces are vastly stronger than gravity, so any net charge would quickly attract opposite charges from the surrounding environment and neutralize itself.

Observational Status of Charged Black Holes

- No charged black hole has been observed.
- Measuring electric charge near a black hole is extremely difficult.
- Current observations allow reliable measurements of **mass** and, in some cases, **spin**, but not charge.

Charged black holes therefore remain a **theoretical possibility**, not an observed reality.

Classification of Black Holes by Mass Scales

Black holes are also classified according to their mass and formation history.

1. Stellar-Mass Black Holes

- **Mass range: ~ 3 to 100 M⊙**
- Formed from the collapse of massive stars
- Most commonly observed black holes

2. Intermediate-Mass Black Holes (IMBHs)

Mass range: 100 to 1 million M⊙

3. Supermassive Black holes(SMBH)

1 million to billions of M⊙

1. Stellar-Mass Black Holes

A stellar-mass black hole is a black hole formed from the collapse of a massive star, typically having a mass ranging from a few to tens of times the mass of the Sun.

Not all stars have the potential to become black holes; only those whose masses fall within specific ranges can collapse into stellar-mass black holes. The final fate of a massive star depends mainly on its initial mass.

Stellar Evolution Ranges

1.1 About 20–50 M⊙: These stars typically collapse to form stellar-mass black holes, either directly or after a supernova explosion.

1.2 About 50–120 M⊙: Some of these stars experience pulsational pair-instability, in which violent eruptions eject large amounts of mass. After losing mass, the remaining core may still collapse into a black hole. In some cases, however, the mass loss is so extreme that no black hole forms.

1.3 About 120–250 M⊙: These stars may undergo pair-instability supernovae, in which the entire star is completely destroyed, leaving no remnant at all; not all massive stars necessarily leave black hole remnants.

Alternative Pathways to Stellar-Mass Black Hole Formation

Stellar-mass black holes can also form indirectly, even when a star does not initially meet the minimum mass requirement.

In **binary star systems**, compact remnants such as **white dwarfs** or **neutron stars** may accrete matter from a companion star. If sufficient mass is accumulated, the object can exceed its stability limit and collapse into a black hole. With sustained mass transfer, they can cross the critical threshold and become a stellar-mass black hole.

Population and Observations

Stellar-mass black holes typically range from a few to several tens of M⊙

Our Milky Way galaxy contains approximately **100 billion stars**, and at least **one** in every **thousand** is massive enough to eventually form a black hole. Based on this estimate, the galaxy may host over **one hundred million stellar-mass black holes**.

This population is not static. As massive stars continue to evolve and die, the number of black holes gradually increases over cosmic time. Because stars evolve at different rates, estimating the exact current number remains difficult.

To date, astronomers have identified **more than 50 stellar-mass black holes** in binary systems. These systems are detectable because matter from a companion star falls onto the black hole, producing intense **X-ray emissions**. Most stellar-mass black holes have been discovered using this method.

A notable example is **Gaia BH3**, a stellar-mass black hole approximately **33 times the mass of the Sun**, currently the most massive known stellar black hole in the Milky Way.

Stellar-mass black holes have also been detected through **gravitational-waves** observations and more than **30 such black holes** have been observed in other galaxies also.

2. Intermediate-Mass Black Holes

Intermediate-mass black holes(IMBHs) occupy the mass range between stellar-mass and supermassive black holes. They typically span from **~100 - 100,000 M⊙.**

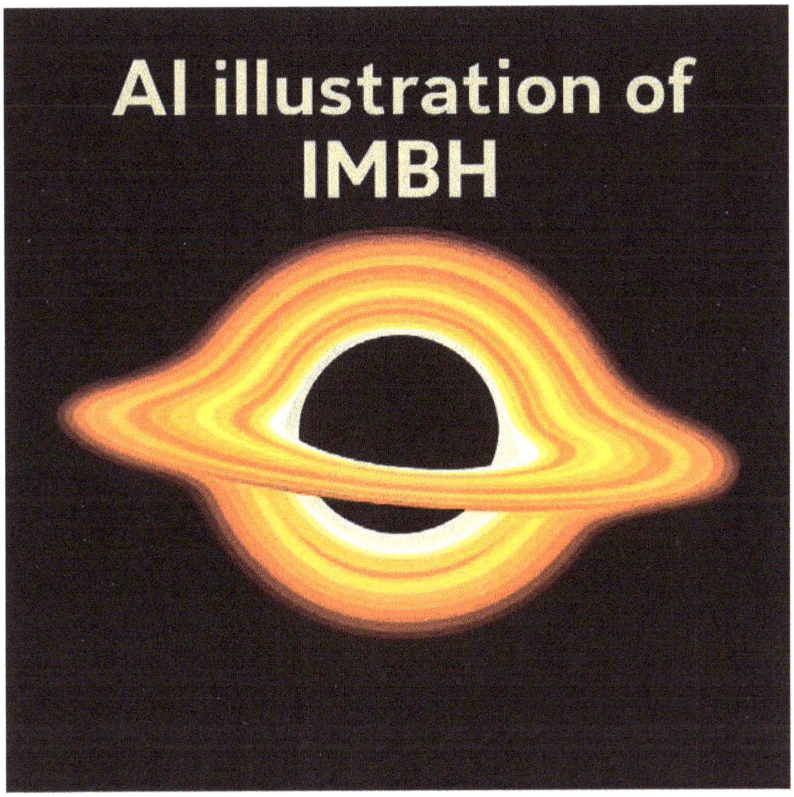

Figure 3.4: AI-generated illustration of IMBH black hole for conceptual purposes

IMBHs are suspected to reside in:

- Dense star clusters,
- Globular clusters,
- Dwarf and small galaxies.

They are often described as the **"missing link"** in black hole evolution, bridging the gap between stellar-mass remnants and supermassive black holes in galaxies.

Detection Challenges

Detecting IMBHs is extremely difficult because they emit very little radiation unless actively accreting matter. Their presence is usually inferred through **gravitational effects**, such as:

- Unusual stellar motions,
- X-ray flares,

- Tidal disruption events.

For example, observations of stellar motions near the centers of globular clusters can indicate the gravitational influence of an unseen massive object.

Studies of the Omega Centauri cluster using the **Hubble Space Telescope** have tracked fast-moving stars that may be bound to an IMBH.

The **Sloan Digital Sky Survey (SDSS)** also identified several X-ray–emitting galaxies hosting IMBHs, some of which show X- ray emission in follow-up observations. However, these detections remain suggestive rather than conclusive.

IMBH Candidate in NGC 4395 Galaxy

In **April 2025**, a team of Indian astronomers made a significant advance by measuring the mass of the black hole at the center of the dwarf spiral galaxy **NGC 4395**, located about **14 million light-years** away from Earth.

Using the **3.6-meter Devasthal Optical Telescope (DOT)** and a technique known as **spectrophotometric reverberation mapping**, they measured the time delay between light from the accretion disk and the response of surrounding gas clouds. The black hole's mass was estimated at approximately **22,000 M⊙ ,** placing it firmly in the IMBH category. This represents one of the more precise mass measurements for an IMBH candidate to date.

Additional observations of the same object ,including X-ray flares and tidal disruption events detected by the XMM-Newton Observatory, suggest the mass of an intermediate-mass black hole (IMBH) of roughly **50,000 M⊙.**

The same object is being studied using different observational techniques, leading to varying mass estimates that are still under investigation; nevertheless, it remains a promising candidate for an isolated intermediate-mass black hole.

Event GW190521: The First Confirmed Intermediate-Mass Black Hole Remnant from the Most Distant Binary Black Hole Merger

Meaning of GW190521

The designation GW190521 refers to a gravitational-wave event detected by the **LIGO and Virgo observatories** on **May 21, 2019.**

The event marked a pivotal moment in astrophysics. It represents the first confirmed observation of an intermediate-mass black hole formed through the merger of two black holes.

Location

The event GW190521 is located about **7 billion light-years** away (light-travel distance), although its present-day distance is much larger—about **17 billion light-years**—due to the expansion of the universe. This makes it one of the most distant confirmed binary black hole mergers, allowing astronomers to probe the universe at an earlier cosmic epoch. The signal lasted for about **0.1 second** within the sensitive band of the detectors.

Black Holes Detection

The signal revealed that two unusually massive black holes of about **85 M⊙** and **66 M⊙** merged to produce a remnant black hole of approximately **142 M⊙,** placing it firmly within the intermediate-mass black hole mass range. Roughly **9 M⊙** of mass was converted into gravitational-wave energy during the merger. This momentarily outshone all the stars in the observable universe combined.

Why This System is Special

GW190521 stands out because it:

1. Challenges existing models of black hole formation
2. Provides strong evidence for intermediate-mass black holes
3. Suggests hierarchical mergers in dense star clusters
4. Binary evolution in dense stellar environments, and possibly early-universe black hole growth
5. The role of repeated mergers in building progressively larger black holes
6. Extends gravitational-wave astronomy to greater cosmic distances
7. It represents the most distant confirmed binary black hole system.

GW190521 strongly supports the idea that **IMBHs can form through hierarchical mergers**, offering an important evolutionary bridge between stellar-mass and supermassive black holes.

Status and Outlook

Although evidence for intermediate-mass black holes (IMBHs) is steadily growing, their true numbers, origins and environments are still uncertain. GW190521 provided the first direct proof that an IMBH can form through a merger, making it a landmark discovery. But merger detections of IMBHs only reveal the end products of such events, not about a **long-lived isolated IMBH**, residing in a galaxy or cluster. Apart from LIGO/Virgo merger events, no isolated IMBH has yet been confirmed beyond doubt. Many candidates exist, but each has alternative explanations.

3. Supermassive Black Holes (SMBHs)

Supermassive black holes (SMBHs) are the largest known class of black holes, with masses ranging from about **one million to several billion times that of the Sun.** Observational evidence strongly indicates that SMBHs reside at the centers of most—if not all—large galaxies. Although it is impossible to count, current estimates suggest that billions, or possibly trillions, exist across the observable universe.

Once considered purely theoretical, SMBHs are now firmly established as common cosmic objects. Their existence is confirmed through the powerful gravitational influence they exert on surrounding stars, gas, dust, and light.

The quasar 3C 273, discovered in the 1960s, provided indirect evidence that supermassive black holes must exist.

Formation of Supermassive Black Holes

The precise origin of SMBHs remains one of the major unsolved problems in astrophysics. Several formation scenarios have been proposed

3.1 Direct Collapse of Supermassive Stars: In the early universe, extremely massive primordial stars may have collapsed directly into black holes without forming supernovae.

3.2 Hierarchical Mergers: Over cosmic time, repeated mergers of stellar-mass black holes may have built progressively larger black holes.

3.3 Accretion of Matter: SMBHs grow by steadily accreting gas, dust, stars, and even other black holes.

3.4 Primordial Black Hole Seeds: Some theories propose that black holes formed shortly after the Big Bang served as "seeds" that later grew rapidly into SMBHs. Advances in observational astronomy continue to refine these models, and future discoveries are expected to clarify the dominant pathways of SMBH formation.

Supermassive black holes are not merely passive objects at the centers of galaxies; they play a key role in shaping their evolution. Their strong gravitational influence governs the motion of nearby stars and gas, while energetic outflows and radiation during active phases can regulate star formation. Observations also reveal a close connection between the mass of a supermassive black hole and the properties of its host galaxy, suggesting that both evolve together over cosmic time. A remarkable example is M87*, the first black hole ever directly imaged.

A Famous Supermassive Black Hole: M87*

Messier 87: A Galactic Giant

Messier 87 is a supergiant elliptical galaxy located about **53-55 million light-years** from Earth in the **Virgo Cluster**. It contains hundreds of billions to about one trillion stars and spans approximately **120,000-132,000 light-years** in **diameter**, making it significantly larger than the Milky Way.

Discovered in **1781** by French astronomer **Charles Messier**, M87 was initially classified as a nebula. Later observations confirmed it as a massive galaxy.

Early Evidence of a Black Hole

In **1917**, astronomers observed **fast-moving, ionized gas** near M87's core, suggesting the presence of an extremely massive compact object. Over time, accumulating evidence confirmed that M87 hosts a supermassive black hole at its center—one of the earliest galaxies known to do so.

Notable Properties of M87*

- **Mass*:* ~ 6.5 billion M⊙**
- **Event Horizon Diameter: ~38 billion kilometers**
- **Spin parameter ≈ 0.9**
- **Magnetic fields: Strong, well-organized magnetic fields**

These magnetic fields play a crucial role in launching and collimating M87*'s powerful relativistic jets.

The Powerful Jet of M87*

One of the most remarkable features of M87* is its powerful relativistic jet, a highly collimated stream of charged particles moving at nearly the speed of light. The jet extends far beyond the galaxy itself and emits radiation across the electromagnetic spectrum—from radio waves to X-rays.

The jet is believed to be powered by the black hole's rotation and magnetic fields, extracting energy from the accretion disk and the spin.

Direct Imaging-A Historic Achievement

The Event Horizon Telescope (EHT)

The Event Horizon Telescope is a global network of radio observatories synchronized to act as a single, **Earth-sized telescope**. It includes facilities such as **ALMA (Chile)**, the

Greenland Telescope, and the **James Clerk Maxwell Telescope (Hawaii)**, among others.

Using **Very Long Baseline Interferometry (VLBI)**, the EHT achieves extraordinary angular resolution, sufficient to image the **shadow of a black hole.**

On **April 10, 2019,** the EHT collaboration released the **first-ever image of a black hole**, capturing the shadow of **M87***.

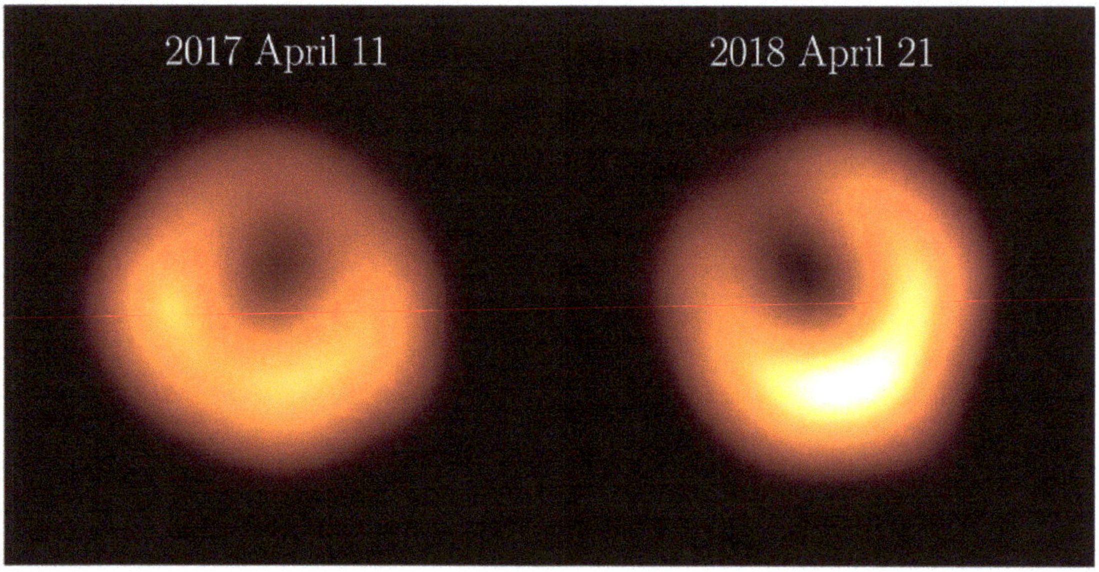

Figure 3.5: The first direct image of a black hole, M87*, obtained by the Event Horizon Telescope in 2019.

Image Credit: Event Horizon Telescope Collaboration

The bright ring represents radiation from hot plasma orbiting near the event horizon, while the dark central region corresponds to the black hole's shadow.

Observations were conducted on April 11, 2017 and the image was released on April 10, 2019. The observed size and shape matched predictions from Einstein's general theory of relativity, providing the first direct visual confirmation of a black hole.

Scientific Significance of the Image

The EHT images allow astronomers to determine:

1. The black hole's mass
2. The size of its event horizon
3. Properties of the surrounding plasma and accretion flow

The first image of M87*, obtained by the Event Horizon Telescope, marked a historic milestone in astrophysics.

Sagittarius A*: The Supermassive Black Hole at the Center of the Milky Way Galaxy

The supermassive black hole Sagittarius A* is located at the center of the Milky Way galaxy approximately **25,000-26,000 light-years** away from Earth. It lies in the direction of the constellation Sagittarius, near the constellations Scorpius and Ophiuchus.

Discovery

In **1974**, American radio astronomers **Bruce Balick** and **Robert L. Brown** discovered a compact **radio source** near the Galactic Center using the **Green Bank Interferometer**, naming it Sagittarius A*.

In the **1990s**, astronomers began tracking stars orbiting this invisible object. These stars exhibited extremely fast, tightly bound orbits, implying the presence of a massive compact object.

A landmark study led by **Reinhard Genzel** and **Andrea Ghez** focused on the star **S2**, which orbits Sagittarius A* in approximately **16 years** and approaches within **16-17 light-hours** from the galactic center.

In **2018**, **S2** reached speeds of **7,650-8,000 km/s** near its closest approach. From these measurements, the mass of Sagittarius A* was calculated to be approximately **4.3 million M⊙** ,with an event horizon radius of about **12-13 million kilometers**.

Additional confirmation came from observations of **gravitational redshift**, a key prediction of general relativity. Alternative explanations—such as clusters of stellar remnants—were ruled out.

"**Reinhard Genzel** and **Andrea Ghez** were awarded the **Nobel Prize in Physics (2020)** for their observations, which provided the first direct and compelling evidence for a supermassive black hole—Sagittarius A*—at the center of the Milky Way.

Transition from M87* to Sagittarius A*

Encouraged by the success of imaging M87*, the EHT turned its attention to a closer but far more challenging target: **Sagittarius A***.

Unlike M87*, Sagittarius A* lies in a crowded, and dust-obscured region, making observations far more difficult. Nevertheless, imaging it offered an unprecedented view of the heart of the Milky Way galaxy.

Direct Imaging of Sagittarius A*

On **May 12, 2022**, the EHT collaboration released the **first image of Sagittarius A***. The image revealed a bright ring surrounding a dark shadow, consistent with a black hole of the predicted mass.

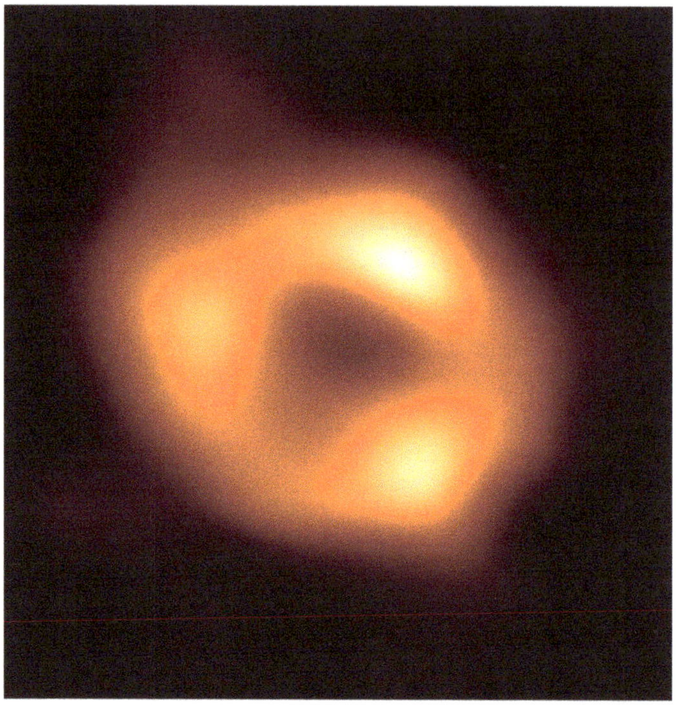

Figure 3.6: Image of Sagittarius A*

Image Credit: Event Horizon Telescope Collaboration

Although Sagittarius A* is much smaller than M87*. Its shadow appears similar in size due to its proximity—providing a striking confirmation of black hole physics across vastly different mass scales.

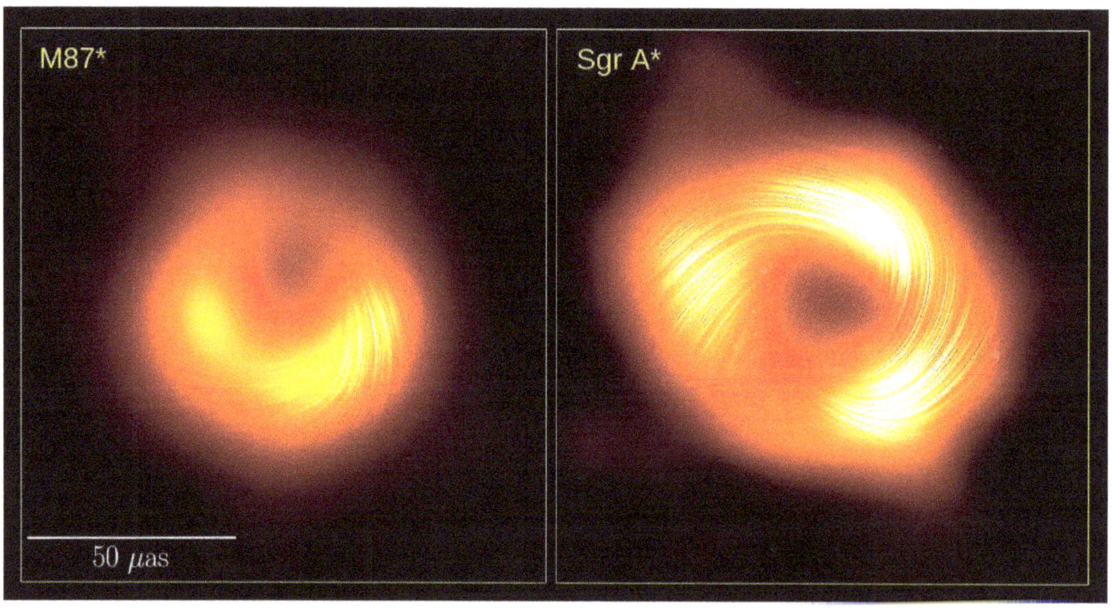

Figure 3.7: Comparison of size of M87* and SagA*

Image Credit: Event Horizon Telescope Collaboration

Recent Unusual Activity Around Sagittarius A*

Although generally dormant, Sagittarius A* occasionally exhibits flaring activity:

January 5, 2015: An X-ray flare~400 times brighter than normal, possibly caused by a disrupted asteroid or gas cloud

May 13, 2019: Sudden infrared brightening observed by the Keck Observatory

June 2023: Discovery of hundreds of mysterious magnetic filaments near the Galactic Center

These events suggest complex magnetic and plasma processes occurring near the black hole.

Nature and Properties of Sagittarius A*

Dormant State: Sagittarius A* accretes matter at a rate **millions of times below** its theoretical maximum (the Eddington limit). As a result, it emits very little radiation. One astrophysicist famously compared this to *"a human eating a single grain of rice every ten million years."* This low activity explains why the Milky Way's core is relatively calm compared to galaxies hosting active quasars.

The Spin of Sagittarius A*: Supermassive black hole Sagittarius A* is spinning much faster—about **60%** of the maximum speed allowed by the laws of physics. Such a high spin hints that the black hole has been actively feeding on surrounding matter over billions of years, gaining energy and momentum as our galaxy evolved. But at present ,it is reremarkably quiet today, accreting very little matter from its surroundings.

This high spin tells a different story about its past. Long ago, Sagittarius A* must have gone through a much more active phase, devouring vast amounts of gas and dust or merging with other massive objects. Its present calm hides the memory of a turbulent youth when it fed greedily and gained the rotation it still carries today.

With a mass of about **4.3 million M⊙,** Sagittarius A* is indeed supermassive, yet modest compared to others—such as the black hole in the galaxy **Holmberg 15A,** located about **700 million light-years** away in the galaxy **cluster Abell 85,** which hosts a supermassive black hole with an estimated mass of roughly **22 billion solar masses (M⊙),** placing it among the most massive black holes known.

The Milky Way is not classified as an Active Galactic Nucleus (AGN) or a Seyfert galaxy, because Sagittarius A* emits very little radiation. A Seyfert galaxy, named after the astronomer Carl Seyfert, contains a supermassive black hole that actively accretes matter, producing intense radiation across the entire electromagnetic spectrum. The infalling gas and dust heat up to millions of degrees, outshining the combined light of billions of stars.

By contrast, Sagittarius A* remains tranquil, allowing life in the Milky Way to evolve under comparatively stable cosmic conditions. This peaceful central black hole may, in a way, be a blessing for our galaxy.

The Solar System's Galactic Journey

Our Solar System lies about **26,000 light-years** from Sagittarius A*. Bound by the galaxy's gravity, the Sun and its planets orbit the galactic center at roughly **230 km/s,** completing **one revolution every 225–250 million years,** known as **a cosmic year**. Since its formation about **4.6 billion years** ago, the Solar System has completed only around **twenty such orbits,** reminding us that Earth, the Sun, and humanity itself are part of a vast, ongoing galactic journey shaped by the gravitational structure of the Milky Way.

Example of Extremely Massive Black Holes

Astronomers estimate that billions of supermassive black holes exist in the observable universe, yet only a small fraction have been discovered and studied in detail. The following are a few remarkable examples that reveal the immense diversity and scale of these cosmic giants:

Phoenix A, with an estimated mass of about **100 billion M⊙,** is currently considered one of the most massive black holes ever proposed, although this value remains under scientific review.

TON 618 is one of the most massive and luminous quasars known, hosting an ultramassive black hole with an estimated mass of about **66 billion M⊙**. Black holes with masses exceeding 10 billion M⊙ are often referred to as ultramassive.

IC 1101 is one of the largest elliptical galaxies, located in the galaxy cluster Abell 2029 in the constellation Virgo, approximately **1.04 billion light-years** from Earth.

Based on empirical scaling relations between galaxy properties and black hole mass, IC 1101 is believed to host an extremely massive central black hole. Some estimates suggest a possible mass in the range of **50–70 billion M⊙,** potentially placing it in the category of an ultramassive black hole.

NGC 4889, located in the Coma Cluster approximately **300 million light-years** from Earth, hosts a central supermassive black hole with an estimated mass of about **21 billion M⊙.**

NGC 1277: Situated in a **lenticular galaxy** within the **Perseus Cluster,** approximately **220–240 million light-years** from Earth, harbors a central supermassive black hole with an estimated mass, roughly **17 billion M⊙,** depending on dynamical modeling methods.

This mass is unusually large relative to the size and stellar mass of its host galaxy, challenging standard models of galaxy–black hole coevolution.

Each of these titans anchors its galaxy, shaping not only its structure but its very fate—silent architects of the universe, holding cosmic order within the reach of their invisible gravity.

Conclusion: The Spectrum of Black Holes

Once considered theoretical curiosities, black holes are now recognized as essential components of the universe. Ranging from stellar-mass remnants of dying stars to supermassive giants at galactic centers, they reveal gravity in its most extreme form. Classified by mass, spin, and charge—as predicted by general relativity, they connect theory with observations such as accretion disks, jets, X-rays, and gravitational waves. While landmark observations have confirmed many predictions, fundamental questions about their origins and growth remain, ensuring that black holes continue to be among the most active frontiers of modern astrophysics.

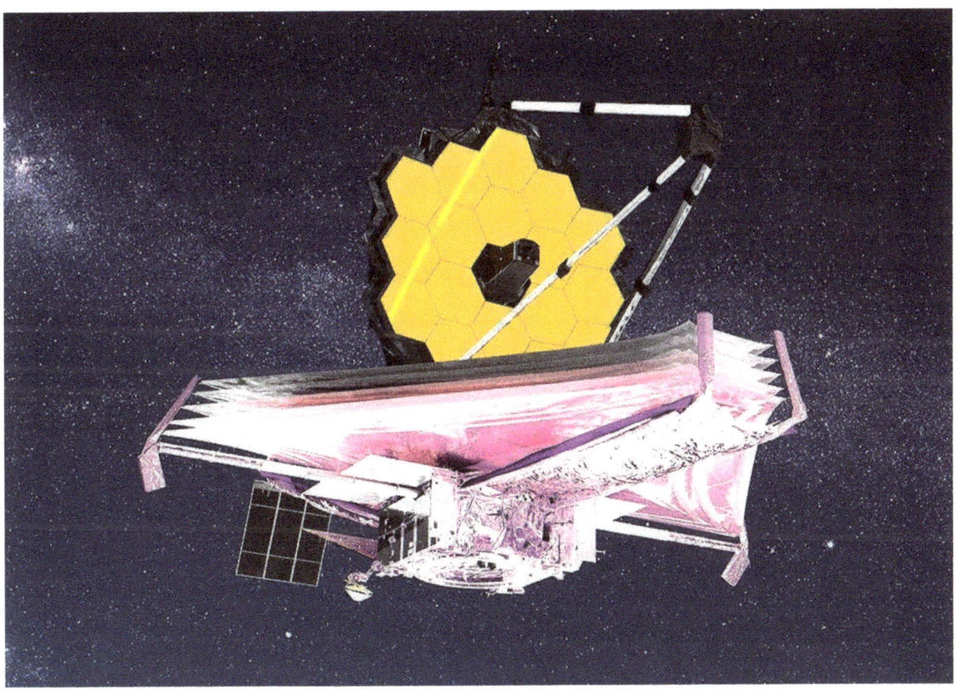

Figure 3.8: The James Webb Space Telescope(JWST)

Image Credit: NASA / Science

It is currently in the constellation of **Taurus** at a distance of about **1550282. 5 km** from earth.

Chapter 4

Black Holes in Binary System

Introduction

In the vast cosmos, many celestial objects do not exist in isolation but instead form **gravitationally bound pairs**. Such systems, known as **binaries**, are among the universe's most fascinating and dynamic structures. They reveal how gravity governs motion, energy exchange, stellar evolution, and including the generation of extreme phenomena such as gravitational waves.

When black holes form binary systems, their interactions transform them into natural cosmic laboratories. These systems allow scientists to probe the deepest laws of physics, test general relativity under extreme conditions, and study how matter behaves in the strongest gravitational fields known in nature.

Binary Systems of Celestial Objects

A **binary system** consists of two celestial objects—such as stars, black holes, neutron stars, or combinations thereof—gravitationally bound and orbiting a common center of mass. Rather than one object simply circling the other, both bodies move in response to their mutual gravitational attraction.

Binary systems are widespread throughout the universe and are not limited to stars alone. Known examples include:

1. **Black hole–star binaries**
2. **Black hole–neutron star binaries**
3. **Black hole–black hole binaries**

1. Black Hole–Star Binary Systems

A black hole–star binary system consists of a black hole and a normal star orbiting a shared center of mass. These systems are among the most important observational tools in black hole astronomy.

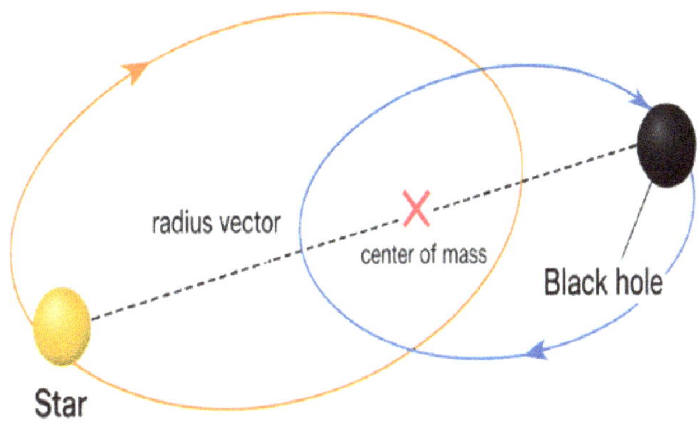

Figure 4.1: Schematic illustration of a black hole- star binary system
(AI-generated; not to scale)

They are typically identified through:

1. X-ray emissions, produced when matter from the companion star is pulled toward the black hole and heated to extreme temperatures in an accretion disk.
2. Orbital motion of the companion star, whose velocity and trajectory reveal the gravitational influence of an unseen massive object.

Earlier we discussed Cygnus X-1 as a binary system. Here are the details of this remarkable object.

1. Black Hole- Star Binary Systems

1.1 Cygnus X-1: First Black Hole

The Cygnus X-1 binary system lies in the constellation Cygnus within our Milky Way galaxy, at a distance of about **6,000 - 7,200 light-years** away from Earth.

It was the first confirmed black hole–companion star binary system, through the pioneering work of **Paul Murdin** and **Louise Webster**, and independently by **C.T. Bolton** in 1971. This discovery provided the first strong observational evidence for the existence of black holes in the universe. Since then, the Cygnus X-1 has become one of the most intensely studied and best-understood black hole systems.

Component of the System

Black Hole: The mass of the black hole in this system is approximately **21 M⊙,** placing it in the category of stellar-mass black holes. It is a rapidly rotating, prograde black hole with a dimensionless spin parameter in the range **0.93 - 0.99**. This corresponds to an extremely fast rotation, with the event horizon effectively completing several hundred revolutions per second. Near the equator, spacetime is dragged around at a significant fraction of the speed of light. Such extreme rotation makes Cygnus X-1 one of the best-known examples of a near-extremal Kerr black hole.

The Schwarzschild radius is approximately **62 km,** while the Kerr outer horizon radius (for high spin) ranges from about **30–60 km,** depending on the spin.

Companion Star: The companion of the black hole in the Cygnus X-1 system is a massive blue supergiant star known as **HDE 226868**. It has a mass of about **20–25 times M⊙,** and an extremely high surface temperature of nearly **31,000 K,** making it intensely bright and bluish. The star shines with a luminosity roughly **300,000–400,000 times greater than that of the Sun**. Its radius is about **15–17 times larger than the Sun's,** showing that it is a true giant star. This remarkable pairing of a small, dense black hole and a huge, brilliant star makes Cygnus X-1 one of the most dramatic and energetic systems in our galaxy.

The black hole and its companion star in the Cygnus X-1 system orbit each other with a period of about **5.6 days**, forming a compact and highly energetic system.

Figure 4.2: Image of black hole in the Cygnus X-1 binary system.

Image credit: NASA/ CXC

Radiation and Jets in Cygnus X-1 System

Cygnus X-1 emits **strong X-rays** because gas from its companion star falls toward the black hole and heats up in the accretion disk. Part of this inflowing matter is also expelled as narrow, **fast-moving jets** that shoot out perpendicular to the disk. These jets, observed mainly in **radio wavelengths**, travel at nearly the speed of light and are most prominent when the system is in its low-energy (hard) state.

Because its stellar-mass black hole produces powerful relativistic jets, the system is classified as a **microquasar.**

Cygnus X-1 is a relatively wide binary system, and current studies indicate that it is unlikely to merge into a single black hole in the future. When the companion star HDE 226868 reaches the end of its life, it is expected to undergo a supernova explosion, forming either a neutron star or another black hole. However, the resulting mass loss and explosion dynamics will most likely disrupt the system rather than lead to a merger.

1.2 LMC X-1: Persistent X-ray

LMC X-1 is a binary system of a stellar-mass black hole and a massive companion star is located in the Large Magellanic Cloud, a satellite galaxy of the Milky Way, around **160,000–165,000 light-years** away from Earth.

The system was first discovered in 1969 during an X-ray survey of the Large Magellanic Cloud using detectors carried on an Aerobee sounding-rocket mission. The nature of the system was later confirmed through spectroscopic studies by **R. A. Hutchings, D. Crampton, and A. P. Cowley in 1983.**

Components of the System

Black Hole: The black hole in the system has a mass in the range of approximately **10.5–11 M⊙.** It is estimated to be a rapidly spinning, prograde stellar-mass black hole, with a dimensionless spin parameter in the range roughly **0.85 - 0.97**. The spacetime at the event horizon is dragged around at an effective equatorial speed of about **1.5×10^8 – $2. 1\times10^9$ m/s,** indicating rapid rotation.

Companion Star: The companion is a hot O-type blue supergiant (spectral type O7–O8) with a mass of about **24–35 M⊙** and a radius roughly **17 times that of the Sun's**. The star's rotational velocity is about **130 km/s.**

The system has an orbital period of approximately **3.9 days**.

X-ray Emission, Nebular Ionization, and Variability in LMC X-1

The powerful X-ray and ultraviolet emission from the system ionizes the surrounding interstellar medium, creating a detectable nebula observable at X-ray and ultraviolet wavelengths. While LMC X-1 does not exhibit prominent, persistent relativistic jets like some other black hole binaries, weak or transient outflows may occur under certain conditions.

Variability in the system is primarily driven by changes in the companion star's stellar wind and in the accretion flow, leading to fluctuations in X-ray output rather than rapid changes in the star's physical size.

Because of its steady X-ray emission, massive companion star, and well-constrained system parameters, LMC X-1 is regarded as one of the most valuable laboratories for studying black hole formation, accretion physics, and the evolution of massive stars.

1.3 M33 X-7: High-mass System

M33 X-7 is a black hole and massive star binary system located in the Triangulum Galaxy (M33), about **2.7 million light-years** away from Earth. It is one of the most distant stellar-mass black hole binaries whose orbital motion has been measured directly. This makes it a true extragalactic X-ray binary.

Components of the System

Black Hole: The M33 X-7 system consists of a stellar-mass black hole of mass around **15.65 ± 1.45 M⊙.** Its dimensionless spin parameter is about **0.77 ± 0. 05.**

Companion Star: The companion star is extraordinarily massive, about **70 M⊙,** hot and highly luminous. The two objects orbit each other every **3.45 days** in a nearly circular, eclipsing orbit, which allows precise measurement of their masses. Matter from the powerful stellar wind of the supergiant is captured by the black hole, forming an accretion disk that emits intense X-rays.

An Unusual High-Mass X-ray Binary: The companion star (~70 M⊙) is among the most massive known in high-mass X-ray binaries, while the black hole (~16 M⊙) is unusually massive for a well-measured stellar-mass black hole. Therefore, the system is considered an **unusually massive high-mass X-ray binary.**

Significance: In typical binary evolution, the initially more massive star evolves first and collapses into a black hole, leaving behind a less massive companion. In M33 X-7, however, the present companion is far more massive than the black hole, indicating complex prior mass transfer and challenging simple stellar evolution models.

Discovery: M33 X-7 was identified as an eclipsing X-ray source through observations made with the **Chandra X-ray Observatory**. Subsequent studies using the **Hubble Space Telescope** and large ground-based telescopes revealed the presence of a massive companion star and allowed astronomers to measure the orbital motion of the system. These observations ultimately confirmed that M33 X-7 is a binary system consisting of a stellar-mass black hole and a massive companion star, a result established in 2007 in a landmark study by **Jerome A. Orosz** and collaborators.

1.4 G3425: Low Mass Black Hole

Circular Orbit

(The binary system **G3425** is also formally identified in astronomical catalogs by its **Gaia ID: 3425577610762832384**)

Location: The system of low black hole and star is located in the Milky Way at a distance of **about 5,800 light-years,** away from Earth. It represents a rare and potentially significant system.

Discovery: The system was identified using data from the Gaia space observatory of the European Space Agency and China's **LAMOST** observatory. The discovery was reported in *Nature Astronomy (2024)* **by Dr. S. Wang** and collaborators**.**

Components of the System

Black Hole: The black hole in the system has an estimated mass in the range **3.1–4.4 M⊙**. **Black holes in this mass range are extremely rare**, as most known stellar-mass black holes lie between 5–25 M⊙. The black hole may have formed via direct collapse (with no strong supernova kick) or through a fallback supernova with minimal asymmetry. This black hole is **dormant** and was identified from the motion of its companion star. Since it has no accretion disk and emits no strong X-rays, its spin parameter cannot currently be measured.

Companion Star: The companion is a red giant star with a mass of approximately **2–3 M⊙.** The system has an **orbital period** of about **880 days** and an almost perfectly **circular orbit.**

Unusual Characteristics and Significance

The wide, circular orbit of G3425 is highly unusual. Typically, the supernova explosion that forms a black hole imparts a powerful "kick," producing an eccentric orbit or disrupting the binary entirely. The survival of G3425 in such a stable configuration challenges current models of supernova physics and binary evolution.

This system provides critical constraints on models of stellar death, compact object formation, and binary survival.

The Growing Population of Black Hole -Star Binaries

Astronomers have identified around fifty black hole–star binary systems within the Milky Way galaxy, along with a smaller number in other galaxies. Once considered rare and extraordinarily difficult to detect, these systems are now recognized as relatively common features of the cosmos, reflecting the growing power of modern observational techniques.

2. Black Hole–Neutron Star Binary Systems

Black hole–neutron star (BH–NS) binaries are among the most extreme and elusive compact systems in the universe. While their mergers have been confirmed through gravitational-wave observations, detecting such systems prior to merger is difficult due to weak electromagnetic signals and the small size of neutron stars. During a merger, the neutron star may be tidally disrupted and produce observable radiation, or it may be swallowed whole, leaving gravitational waves as the only detectable signal.

2.1 Event GW200105 (5 January 2020)

This event involved the merger of a black hole with a mass of about **9 M⊙** and a neutron star of approximately **1.9 M⊙**, located nearly **900 million light-years** from Earth.

The gravitational-wave signal was detected by the LIGO, Virgo and KAGRA detectors at frequencies starting around **20 Hz to over 1000 Hz during final stages of merger.** The final black hole formed after the merger had a mass slightly lower than the combined mass of the two objects, with the difference carried away as gravitational-wave energy. The inferred orbital properties are consistent with both isolated binary evolution and dynamical formation in dense stellar environments.

GW200105 was the first confirmed black hole–neutron star merger, providing direct observational evidence that such mixed compact-object systems exist. The signal captured the inspiral phase during which the black hole and neutron star remained distinct before their final coalescence, offering valuable insight into strong-field gravity and compact binary evolution.

2.2 Event GW200115 (15 January 2020)

GW200115 was the **second confirmed black hole–neutron star merger event**, detected on 15 January 2020 by the LIGO–Virgo Collaboration.

The system consisted of a black hole of about **5.7 solar masses** and a neutron star of approximately **1.5 solar masses,** located at a distance of nearly **1 billion light-years** from Earth. The gravitational-wave signal swept upward in frequency, ranging from about **25 Hz** to **several hundred hertz** before the final merger.

Absence of Electromagnetic Counterparts

Despite extensive follow-up observations, **no electromagnetic counterparts**—such as gamma-ray bursts or kilonovae—were detected for either GW200105 or GW200115. This absence suggests that, in both cases, the neutron star was likely swallowed by the black hole without significant tidal disruption, or that any resulting emission was too weak or unfavorably oriented to be detected.

A Contrasting Case: GW170817

In contrast, the neutron-star merger event **GW170817(17 August, 2017) involved two neutron stars** with a combined binary mass of approximately **2.73–2.74 M⊙**.

The merger occurred in the galaxy **NGC 4993,** located about **130–140 million light-years** from Earth, and produced both gravitational waves and a bright electromagnetic counterpart.

This landmark discovery marked the **first simultaneous detection of gravitational waves and electromagnetic radiation** from the same astrophysical event. It confirmed long-standing theoretical predictions that neutron-star mergers produce kilonovae and demonstrated that such events are an important source of heavy elements in the universe through rapid neutron-capture (r-process) nucleosynthesis.

Kilonovae and Their Origin

A kilonova is a short-lived but powerful astronomical event that can occur when two neutron stars merge, or when a neutron star merges with a black hole under favorable conditions. Kilonovae emit **visible and infrared** regions of the electromagnetic spectrum. Whether a merger produces observable electromagnetic radiation depends critically on the interplay between the black hole's mass, spin and the neutron star's internal structure.

3. Binary System of Stellar-Mass Black Holes

3.1 Event GW150914

The historic event GW150914, detected **on 14 September 2015**, marked the **first direct observation of gravitational waves** and the first confirmed detection of a binary system composed of two stellar-mass black holes. The signal was independently observed by the LIGO detectors at Hanford and Livingston.

This groundbreaking detection was made with extremely high statistical significance, leaving no doubt about its astrophysical origin. In this merger, one black hole had a mass of approximately **36 M⊙**, while the other had a mass of about **29 M⊙**.

The event occurred at a distance of roughly **1–1.3 billion light-years** from Earth and was localized to the southern celestial hemisphere.

Following the merger, the newly formed black hole had a mass of about **62 M⊙** and was rapidly spinning. The remaining **~3 M⊙** were converted directly into gravitational-wave energy, in precise agreement with predictions from Einstein's general theory of relativity. The observable inspiral–merger–ringdown signal lasted approximately **0.2 seconds** within LIGO's sensitive frequency band.

Scientific Significance

The detection of GW150914 represents a landmark achievement in modern physics. It provided the first direct confirmation of gravitational waves, demonstrating that ripples in spacetime produced by accelerating massive objects can be detected.

Although binary black holes had been studied theoretically for decades, this event transformed a long-standing prediction into observational reality and opened an entirely new window on the universe.

3.2 Event GW151226

Just four months later, on **26 December 2015,** the LIGO Scientific Collaboration detected a second binary black hole merger, designated GW151226.

This system consisted of two stellar-mass black holes with masses of approximately **14.2 M⊙ and 7.5 M⊙**, which merged to form a black hole of about **20.5 M⊙.**

The event occurred at a distance of roughly **1.4 billion light-years** from Earth.

Absence of Electromagnetic Counterparts

As expected for binary black hole mergers, no electromagnetic radiation—such as visible light, X-rays, or gamma rays—was detected from GW150914 or GW151226. Unlike neutron star mergers or supernova explosions, black hole mergers are predicted to emit energy almost exclusively in the form of gravitational waves. These observations strongly support that prediction.

Scientific Significance of GW151226

The detection of GW151226 reinforced and expanded the significance of GW150914. Together, these events confirmed that mergers of stellar-mass binary black holes are relatively common in the universe and revealed a diverse population of black holes with varying masses, spins, and orbital properties.

Present Observational Status

Since 2015, the LIGO–Virgo–KAGRA (LVK) collaboration has detected approximately 200 binary black hole merger events through gravitational waves. To date, however, a still-unmerged binary system consisting of two stellar-mass black holes has not been directly observed as a resolved pre-merger system. Such binaries are expected to be rare and extremely difficult to detect, as they emit little or no electromagnetic radiation prior to coalescence.

Gravitational-wave detectors are currently sensitive primarily to the final seconds to minutes of the inspiral and merger phases, when the orbital separation becomes sufficiently small to produce strong spacetime distortions. As a result, binary black hole systems are typically discovered only at the moment of merger rather than during earlier stages of their evolution.

Despite this limitation, the growing catalog of detections has enabled detailed population studies of stellar-mass black holes, revealing a wide range of component masses, mass ratios, and spin configurations. These observations have provided important insights into black hole formation channels, including isolated binary evolution and dynamical assembly in dense stellar environments such as globular clusters and galactic nuclei.

Future upgrades to ground-based detectors, along with next-generation observatories such as the **Einstein Telescope** and **Cosmic Explorer,** are expected to significantly improve sensitivity. These instruments may allow black hole binaries to be observed earlier in their inspiral phase, extending the observable duration of gravitational-wave signals and further refining constraints on their astrophysical origins.

4. Binary System of Intermediate-Mass Black Holes

Event GW231123

The gravitational-wave event **GW231123**, detected on **23 November 2023** by the **LIGO–Virgo–KAGRA (LVK) collaboration**, provided the first strong observational evidence for a binary system composed of two intermediate-mass.

This remarkable event involved the merger of black holes with masses of approximately **135 M⊙** and **103 M⊙**, placing both objects firmly within the intermediate-mass black hole regime. At the time of detection, this was among the **most massive black hole merger events ever observed** through gravitational waves. Both black holes exhibited **exceptionally high spin rates**, with the larger object having a dimensionless spin parameter of approximately **0.9**, and the smaller approximately **0.8**.

Figure 4.3: Conceptual illustration of a binary IMBH system during inspiral (AI-generated, not to scale)

Their coalescence produced a final black hole with a mass of roughly **225 M⊙**, forming a rapidly spinning **intermediate-mass black hole**. The entire merger unfolded in an extraordinarily short interval of about **0. 1 seconds**, emitting gravitational waves in the frequency range of approximately **30–80 Hz**.

The source was located at an estimated distance about **2. 3–13. 4 billion light-years** away from Earth , making it one of the **most distant black hole mergers** detected to date.

Scientific Importance

While the detection of the event is not any doubt, there is scientific debate and doubt regarding the exact nature of the progenitor stars whether they were formed by traditional stellar evolution or other mechanisms.

GW231123 provided strong evidence that intermediate-mass black hole binaries exist, confirming long-standing theoretical predictions. The event offers insight into possible formation pathways, such as repeated mergers in dense stellar environments, while more speculative primordial origins cannot be ruled out. The system remained invisible to conventional telescopes and revealed itself only during its final merger through gravitational waves, highlighting the unique power of gravitational-wave astronomy to uncover hidden black hole populations.

Challenges in the Detection of Intermediate-Mass Black Hole Binaries

Intermediate-mass black hole (IMBH) binaries are extremely hard to observe before merger because they emit little or no light. They often lack nearby gas or stars to accrete, and when they form in dense environments such as star clusters, their subtle gravitational effects are easily hidden among surrounding stars. Even when weak emission is present, it is difficult to distinguish from background sources. As a result, IMBH binaries usually reveal themselves only during the brief final moments of merger, when gravitational waves become detectable, after spending most of their lives in silence.

From IMBHs to Supermassive Black Hole Binaries

While gravitational-wave detections have confirmed the existence of merging black hole binaries, it is important to note that **not all black hole pairs merge within the age of the universe**. This is especially true for systems involving **supermassive black holes (SMBHs)**, which may orbit each other for billions of years without coalescing.

As a result, astronomers have begun searching for **long-lived, non-merging binary black hole systems**, particularly at the centers of galaxies, where supermassive black holes are expected to form pairs following galaxy mergers. Such systems are thought to form naturally when galaxies merge and their central black holes move toward a common center.

Despite strong theoretical expectations, confirming binary SMBH systems has proven extraordinarily difficult. Although astronomers have identified several promising candidates exhibiting periodic light variations, jet distortions, or dual active nuclei, no supermassive black hole binary has yet been observed in a manner that is universally accepted as definitive.

Among all proposed candidates, however, **OJ 287** stands out as the most compelling and extensively studied example.

5. The Binary Supermassive Black Hole System

5.1. OJ 287: A Compelling SMBH Binary System

OJ 287 is one of the most remarkable extragalactic systems and is widely regarded as one of the most compelling candidates for a binary supermassive black hole system.

The binary system of supermassive black holes in OJ 287 was proposed by Finnish astronomer **Aimo Sillanpää** and his associates at the University of Turku, who first published the hypothesis in **1988**.

The object OJ 287 was first identified as a variable radio source in 1967 during radio surveys of the sky conducted by astronomers studying extragalactic radio emitters. It was later recognized as a **BL Lacertae–type active galactic nucleus,** producing a powerful jet directed nearly toward Earth, showing strong variability in optical and radio wavelengths. Over time, long-term monitoring of its brightness revealed a remarkable pattern of nearly periodic optical outbursts occurring roughly every 12 years.

If a galaxy shows bright outbursts at regular intervals—such as every 12 years—it suggests that something is repeating in a predictable cycle. One natural explanation is the presence of two supermassive black holes orbiting each other. As the smaller black hole moves along its orbit, it can periodically pass through the disk of gas surrounding the larger black hole. Each time this happens, it disturbs the gas and produces a burst of radiation, which we observe as a flare or outburst. The regular timing of these events reflects the orbital period of the binary system.

These unusual eruptions eventually led astronomers to propose that OJ 287 contains a binary system of supermassive black holes, although this interpretation emerged only after decades of observational study.

Black Holes in the System

OJ 287 is modeled as containing two supermassive black holes of dramatically different masses:

Primary black hole: approximately **1.8×10^{10} M⊙,** with a dimensionless spin parameter estimated to be **0.38** based on relativistic orbital modeling.

Secondary black hole: approximately **1.5×10^8 M⊙**.

The supermassive black hole binary system OJ 287 has an **orbital period** of approximately **12 years.**

Location and Observational Properties

OJ 287 lies roughly 3.5–4 billion light-years from Earth in the constellation **Cancer**. Its optical brightness has been monitored for more than **130 years**, making it one of the longest-studied **active galactic nuclei.**

The system produces a powerful relativistic jet oriented nearly along our line of sight. This geometry causes relativistic beaming, in which even small intrinsic variations in emission appear strongly amplified. For this reason, OJ 287 is classified as a **blazar**.

Why OJ 287 is Considered a Binary Supermassive Black Hole Candidate

OJ 287 shows several key features that suggest it contains two supermassive black holes:

1. **~12-year quasi-periodicity:** The system exhibits a repeating cycle of major optical outbursts roughly every 12 years, consistent with models in which a secondary supermassive black hole orbits a much more massive primary.
2. **Disk-impact flares:** In the leading model, the secondary black hole is predicted to plunge through the accretion disk of the primary at relativistic speeds (estimated on the order of 20,000–25,000 km/s), producing intense, short-lived optical flares.
3. **Predicted outburst timing:** Several major flares have occurred at times broadly consistent with predictions from relativistic binary orbital models, including effects such as orbital precession.
4. **Extreme luminosity:** During major outbursts, OJ 287 can appear brighter than an entire galaxy like the Milky Way due to relativistic beaming, reaching luminosities exceeding a **trillion times that of the Sun.**
5. **Orbital evolution:** Modeling suggests a gradual decrease in orbital separation consistent with the level of energy loss expected from gravitational-wave emission in a relativistic binary system, with a merger predicted in approximately **10,000 years,** although gravitational waves from OJ 287 have not yet been directly detected.
6. **High-energy emission:** Gamma-ray flares observed by NASA's Fermi Gamma-ray Space Telescope provide additional insight into jet activity and may be related to processes occurring during disk-impact events.

Together, these independent lines of evidence make OJ 287 one of the most compelling and extensively studied candidates for a binary supermassive black hole system, although it is not yet directly resolved into two separate black holes.

Limits of Confirmation: Despite strong evidence, OJ 287 is not yet definitively confirmed as a binary system for several reasons:

1. **No direct imaging:** The two black holes have not been clearly resolved as separate objects; the binary nature is inferred from models rather than direct observation.
2. **Model dependence:** The~12-year outbursts fit the binary model, but other explanations—such as jet motion or disk instabilities—may also produce similar variability.
3. **No direct gravitational-wave signal:** A binary system should emit gravitational waves, but no individual signal from OJ 287 has yet been detected.
4. **Variable high-energy emission:** X-ray and gamma-ray activity is irregular and can arise from jet processes alone, not necessarily from a binary system.

For these reasons, OJ 287 remains one of the strongest and most extensively modeled candidates for a binary supermassive black hole system, though it has not yet been definitively confirmed.

Other Prominent Candidate System:

5.2. 4C+37. 11

The radio galaxy 4C+37. 11 (also known as B2 0402+379), located about **750 million light-years** away from Earth, hosts a binary supermassive black hole system at its center.

The system contains two supermassive black holes with a combined mass of roughly **15 billion M⊙.** Their separation is about **7 light-years,** making it one of the smallest directly resolved separations between two supermassive black holes.

The orbital period is estimated to be on the order of **~30,000 years**, although this depends on the assumed orbital parameters.

The binary supermassive black hole system 4C+37. 11 was identified in 2006 by a team led by **Greg B. Taylor** using high-resolution radio observations with very long baseline interferometry (VLBI).

Evidence for a SMBH Binary System: 4C+37. 11 is considered one of the strongest cases of direct spatial evidence for a supermassive black hole binary because:

1. Two distinct compact radio cores have been directly imaged using VLBI.
2. Their separation (~7 light-years) is among the smallest ever resolved for such systems.

3. Both cores show active radio emission, indicating that each likely hosts an accreting supermassive black hole rather than being jet features.
4. Their relative motion is consistent with a gravitationally bound system, though precise orbital parameters remain uncertain.
5. The system likely formed through a galaxy merger, consistent with models of galaxy evolution.

Limits of confirmation: Although two compact cores are clearly observed in 4C+37. 11, the extremely slow orbital motion and lack of precise dynamical measurements mean that the system, while strongly indicative of a binary, is not yet definitively confirmed.

If fully confirmed dynamically, 4C+37. 11 would represent one of the closest directly resolved supermassive black hole binaries known and provides some of the strongest spatial evidence for such systems.

The Challenge of Confirming SMBH Binaries

Confirming supermassive black hole binaries is difficult because their **orbital periods** are **extremely long,** their **separations are hard to resolve** even with the best telescopes, and most evidence is indirect. In addition, many observed signals can also be explained by other astrophysical processes, making clear and unambiguous identification challenging.

Conclusion of the Chapter

Black holes in binary systems reveal their nature most clearly through interaction. Whether paired with a normal star, a neutron star, or another black hole. These systems expose extreme gravity through accretion, radiation, and gravitational waves. Stellar-mass binaries demonstrate how black holes form and evolve, while black hole–black hole mergers, detected through gravitational waves, have opened a new observational window on the universe. At the largest scales, the supermassive binary system OJ 287 extends this picture to galactic dimensions, showing that binaries are fundamental laboratories for understanding black holes across cosmic scales.

Chapter 5
Extreme Cosmic Systems, Objects and Limits

Introduction

In the previous chapters, we explored how black holes form and how they behave within binary systems. Building on that foundation, this chapter expands the discussion to the most extreme systems and objects known in the universe—where gravity, energy, and matter operate at their limits.

Here we examine remarkable cosmic systems such as **V404 Cygni,** the **Sagittarius A*–D9 system,** and rare multiple–black-hole configurations, along with unusual cases like **runaway** and **rogue black holes**, which travel through space after being ejected from their original environments. Recent observations in **December 2025** have revealed rare systems in which **three supermassive black holes,** each associated with a **different galaxy,** are gravitationally bound during ongoing galaxy mergers. The chapter also explores powerful cosmic sources including **quasars** and **blazars,** whose enormous energy output makes them visible across billions of light-years.

The chapter briefly addresses the speculative idea of **white holes,** included as a theoretical concept that highlights the frontiers and open questions of modern gravitational physics.

To understand how such objects shine and evolve, key physical ideas such as the **Eddington limit** are discussed, along with the concept of **cosmological coupling,** which explores possible connections between black holes and the large-scale evolution of the universe.

Together, these topics show how extreme cosmic systems act as natural laboratories for testing gravity, high-energy physics, and our understanding of the universe at its most powerful and mysterious scales.

1. V404 Cygni: A Remarkable Triple System

The V404 Cygni system is an extraordinary celestial arrangement consisting of a stellar-mass black hole and a companion star locked in a close binary system accompanied by a third star orbiting this inner pair. This rare configuration provides the first observational evidence of a black hole–star binary embedded within a triple-star system, making V404 Cygni a unique laboratory for studying complex gravitational interactions and black hole formation.

V404 Cygni is located about **7,800 light-years** away from Earth in the constellation Cygnus, within the Milky Way.

The system's triple nature was reported in 2024 by **Kevin Burdge** and collaborators at MIT and Caltech, based on precise measurements from the Gaia mission of the European Space Agency.

System Configuration

Inner Binary System

The inner system consists of a stellar-mass black hole and a companion star.

Black Hole: The black hole has a mass of about **9–10 M⊙.** Evidence suggests it likely formed through **direct collapse**, without a strong supernova explosion. Such a quiet formation would help the system remain intact, since powerful supernova "kicks" can disrupt nearby stars.

Companion Star: The companion is a **K-type star** with a mass of about **0.7 M⊙** and a radius roughly **6 times that of the Sun.** Its surface temperature is around **4,800 K**, and it rotates rapidly, with an equatorial velocity of about **36 km/s.**

Orbital Period: The two objects orbit each other every **6.47 days.** Because of their close separation, material from the star flows toward the black hole, forming an **accretion disk** that produces energetic outbursts.

Outer (Tertiary) Star

A third star orbits the inner binary at a large distance of about ≈ **5.24×10^{11} km.** This **G-type star** has a mass of about **1.8–1.9 M⊙** and a radius roughly **1.8–2 times that of the Sun.** Its surface temperature is near **6,100 K,** and it completes one orbit in about **70,000 years.** Although only weakly bound, its long-term stability confirms that V404 Cygni is a **hierarchical triple system.**

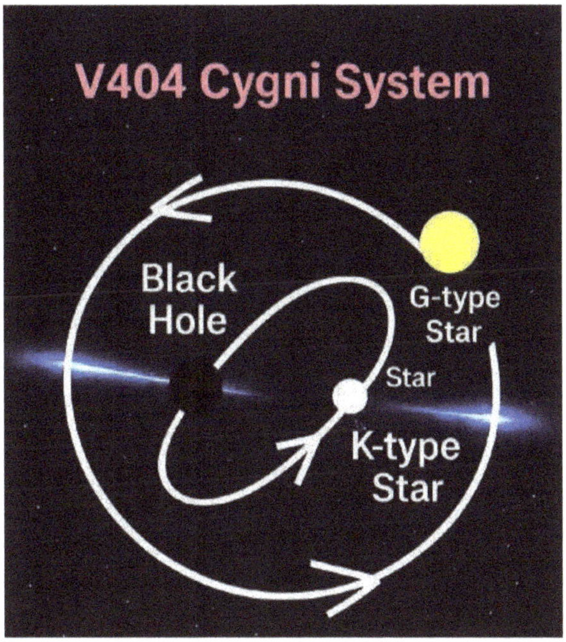

Figure 5.1: Artist-generated schematic illustration of V404 Cygni System

Jet Activity and Microquasar Behavior

The black hole in the V404 Cygni system is a well-known microquasar, producing relativistic jets—streams of particles launched at near-light speed. These jets can show rapid changes in direction, or precession, on very short timescales, sometimes within minutes during active phases.

This behavior likely arises from complex interactions between the black hole's spin, the accretion disk, and the binary system's orbital motion, providing a valuable opportunity to study jet physics and relativistic effects on a stellar scale.

Figure 5.2: X- ray rings around V404 Cygni

Image Credit: NASA/CXC/Univ. of Wisconsin–Madison/S. Heinz et al.

X-ray Rings Around V404 Cygni

A spectacular set of X-ray rings around V404 Cygni was observed by NASA's Chandra X-ray Observatory. These rings are produced when X-rays from a bright outburst are scattered by interstellar dust clouds along the line of sight. They provide valuable information about the distribution of dust and the structure of the space between V404 Cygni and Earth.

Formation, Evolution, and Significance

The survival of a triple system containing a black hole is highly unusual. With an estimated age of about **3–4 billion years,** V404 Cygni demonstrates remarkable long-term stability.

Evidence suggests that the black hole likely formed through a weak supernova or direct collapse, producing only a small "kick," which allowed the system to remain bound.

Because of its compact inner binary, distant third star, and occasional jet activity, V404 Cygni serves as an exceptional natural laboratory. It helps astronomers study black hole formation, accretion processes, jet production, and the evolution of complex gravitational systems.

2. Sagittarius A* – D9 System: A Binary system of Star Orbiting SMBH

The Sagittarius A*– The D9 system represents an extraordinary gravitational arrangement at the very center of our galaxy. It is the first confirmed known system in which a binary pair of stars orbits a supermassive black hole, revealing the delicate balance of motion and gravity in one of the most extreme environments in the universe.

In December 2024, an international team of astronomers led by **Florian Peißker** of the University of Cologne, Germany, reported the discovery of D9 as the first confirmed binary star system orbiting Sagittarius A*. The discovery was made using the Very Large Telescope of the European Southern Observatory, based on more **than 15 years of observations.** The results were published in *Nature Communications*.

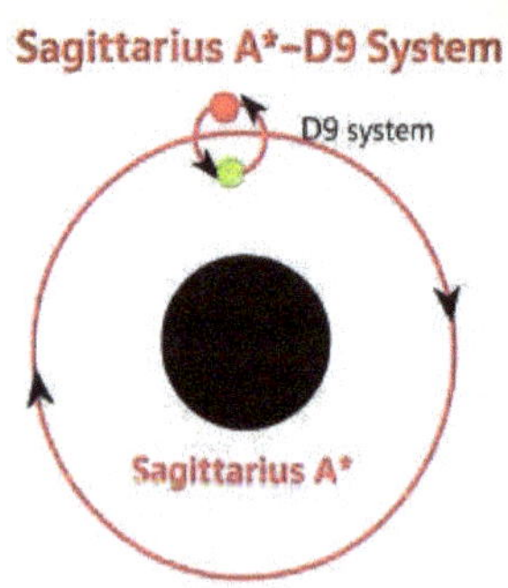

Figure 5.3: Artist's generated illustration of a Sag A*-D9 system

The D9 binary system lies approximately **0.02 light-years** from Sagittarius A* and about **27,000 light-years** from Earth. It is part of the **S-cluster**, a dense group of stars surrounding the Galactic center.

Properties of the Binary Star System

The D9 system consists of two stars:

Primary star: about **2.8 M⊙**

Companion star: about **0.7 M⊙**

The two stars orbit their common center of mass with a period of about **372 days**.

Orbit Around Sagittarius A*

What makes this system exceptional is that the entire binary also orbits Sagittarius A*, resulting in two simultaneous motions:

1. The stars orbit each other
2. Together, the binary orbits Sagittarius A* at a velocity of about **680 km/s**, completing **one orbit** in roughly **360 years**

This dual motion offers a rare opportunity to study gravitational interactions near a supermassive black hole.

Young and Fragile System

D9 is estimated to be only about **2–3 million** years old, making its survival in such an extreme environment remarkable. Strong tidal forces, intense radiation, and frequent stellar encounters threaten its stability. Models suggest that the two stars may eventually merge into a single star within about **a million years,** likely driven by tidal interactions and orbital evolution in the gravitational field of the supermassive black hole.

Scientific Significance

The Sagittarius A–D9 system stands as a unique natural laboratory. It provides direct insight into star formation, stellar survival, and gravitational dynamics in the immediate vicinity of a supermassive black hole—an environment once thought too extreme for binary stars to exist.

3. A Unique Triple System of Supermassive Black Holes: J1218+2953/ J1219+1035

In 2025, December, astronomers reported the discovery of a remarkable and rare cosmic system consisting of **three gravitationally bound galaxies,** each hosting an **actively accreting supermassive black hole** at its center. This extraordinary discovery represents one of the most unusual and important findings in extragalactic astronomy in recent years.

Figure 5.4: AI-generated illustration of Triple AGN system for explanatory purposes

A team of astronomers led by **Dr. Emma Schwartzman** of the U. S. Naval Research Laboratory identified an exceptionally rare triple system of actively feeding supermassive black holes within a group of merging galaxies in the J1218+2953 / J1219+1035 field. High-resolution radio observations using the National Science Foundation's Very Large Array (VLA) and the Very Long Baseline Array (VLBA) revealed compact radio cores in all three galactic nuclei. The results were published in *The Astrophysical Journal Letters (2025).*

These observations confirm that each nucleus hosts an actively accreting supermassive black hole. All three exhibit strong radio emission—likely associated with relativistic jets—making this system one of the first known examples in which three interacting galaxies simultaneously host **radio-bright active galactic nuclei (AGN).**

The system lies at a distance of approximately **one billion light-years** from Earth. The three galaxies are in the process of gravitational interaction and early-stage merging, with projected separations of roughly **22,000** and **27,000 light-years**. Prominent tidal features observed in the system provide clear evidence of strong dynamical interaction.

Scientific Significance: This triple system provides a unique natural laboratory for studying how galaxy interactions trigger black hole accretion, enhance nuclear activity, and drive galaxy evolution. Systems of this kind are also important precursors to future black hole mergers, which may eventually produce low-frequency gravitational waves detectable by next-generation observatories.

Because the system appears to be in an early stage of interaction, continued observations are expected to yield valuable insights into the formation and evolution of multi–black-hole systems over cosmic time.

4. NGC 6240: A Candidate Triple Supermassive Black Hole System

The galaxy NGC 6240, often called the Butterfly Galaxy or Starfish Galaxy, is a highly disturbed merging system located approximately **320–400 million light-years** from Earth in the constellation **Ophiuchus.**

Characteristics: Formed from the collision of two massive galaxies, it exhibits a chaotic structure, intense starburst activity, and powerful infrared emission, making it one of the most extensively studied galaxy mergers in the nearby universe.

Discovery: For many years, NGC 6240 was known to host two actively accreting supermassive black holes, one in each of its merging galactic nuclei. These were confirmed through X-ray and infrared observations, particularly by the Chandra X-ray Observatory, which revealed two compact and luminous sources separated by about **1 kiloparsec (~3,000 light-years).**

In 2019, a detailed spectroscopic study led by **Wolfram Kollatschny**, using the MUSE integral-field spectrograph on the Very Large Telescope, provided evidence that the southern nucleus may itself contain two distinct supermassive black holes. This raised the possibility that NGC 6240 hosts three SMBHs, making it a candidate triple system.

System Configuration: The system is currently interpreted as follows:

Northern SMBH: ~ (8–15) × 10^8 M⊙

Southern SMBH (primary): ~ (7–10) × 10^8 M⊙

Southern SMBH (candidate companion): ~ (5–8) × 10^8 M⊙

The two main galactic nuclei are separated by about **1 kiloparsec**, while the possible third black hole lies within the southern nucleus, making the system dynamically complex.

Scientific Significance: The system is still under investigation and is a promising candidate. If confirmed, NGC 6240 would represent one of the clearest nearby examples of a triple supermassive black hole system formed through hierarchical galaxy mergers. Such systems are of great importance because:

1. They provide direct insight into how galaxies grow through repeated mergers.
2. They help explain how supermassive black holes form close binaries and eventually merge.

3. They are potential precursors to gravitational-wave events, especially in the low-frequency regime detectable by future space-based observatories.

Because NGC 6240 is relatively nearby, it offers a rare opportunity to observe black hole interaction and merger dynamics in progress, bridging the gap between theoretical models and observable reality.

Cosmic Objects Powered by Black Holes

1. Quasars: Beacons of the Early Universe

The Brightest Lights in the Cosmos

Quasars are among the most powerful and distant objects ever observed. These cosmic beacons shine so intensely that they can outshine entire galaxies, allowing astronomers to study the universe as it existed billions of years ago. Found at the centers of massive galaxies, quasars mark periods when their central black holes were actively growing and releasing enormous energy.

Brief History of Discovery

Quasars were first identified in the **1950s** as powerful radio sources that appeared star-like in optical images, making their true nature initially unclear. A major breakthrough came in **1963,** when astronomer **Maarten Schmidt** analyzed the spectrum of the radio source 3C 273 and discovered an unusually large redshift. This demonstrated that the object lay at a vast cosmological distance and possessed an extraordinary luminosity.

The term "quasar"—short for quasi-stellar radio source—was introduced in **1964** by **Hong-Yee Chiu.** The recognition of quasars as distant, highly energetic objects provided strong observational support for an expanding universe and was consistent with predictions of the Big Bang theory. Since then, large sky surveys, most notably the **Sloan Digital Sky Survey**, have cataloged approximately **750,000–800,000 spectroscopically** confirmed quasars, establishing them as key probes of the early universe and cosmic evolution.

Extraordinary Characteristics of Quasars

1. Extraordinary Brightness of Quasars

The extraordinary brightness of quasars does not come from the black hole itself, but from the hot, rapidly rotating accretion disk surrounding a supermassive black hole. As gas and dust fall inward, gravity compresses and heats the material to millions to hundreds of millions of degrees, causing it to emit radiation across the entire electromagnetic spectrum — from radio waves to gamma rays. The highest-energy emission arising from the disk's hot corona and, in many cases, from relativistic jets.

Quasars appear point-like in telescopes because their energy is released from a very compact region at the galaxy's center, even though they can outshine their entire host galaxies.

2. Efficient Energy Conversion

The energy released during accretion onto a black hole is extraordinarily high. In ordinary stars, nuclear fusion converts only about **0.7%** of mass into energy, but when matter falls toward a black hole, about **10%** — and in the theoretical limit of a **rapidly spinning (Kerr) black hole up to nearly 40%, of the infalling mass energy can be converted into radiation.** This happens not through **nuclear reactions**, but because the intense **gravitational pull** of the black hole releases enormous **gravitational energy** as matter spirals inward and heats up in the accretion disk.

Because of this highly efficient conversion of mass into energy, quasars are among the most powerful energy-producing systems in the universe.

3. Relativistic Jets of Quasars

Many quasars launch **twin jets of plasma** that shoot outward from the black hole's poles at nearly the speed of light. These jets can extend far beyond the host galaxy and are visible across vast cosmic distances, especially in radio wavelengths.

Jet formation is driven by the interaction between the black hole's rotation, the accretion disk, and intense magnetic fields. In some cases, a small fraction of the jet's energy may be extracted directly from the black hole's spin, but most originates in the accretion disk itself. Unlike stars, which radiate evenly in all directions, quasars release energy in highly focused beams.

Figure 5.5: AI-generated illustration of quasar for explanatory purposes

4. High Redshift

Quasars exhibit cosmological redshifts (z) ranging from low values to over $z \approx 7$, indicating that they are observed across a vast span of cosmic history—from relatively recent epochs to the early universe.

5. Compact Powerhouses

Their immense luminosity, originating from an extremely small region, points to gravity-driven energy release—specifically, gravitational accretion onto a compact massive object.

6. Broad Spectrum Emission

Quasars emit radiation across nearly the entire electromagnetic spectrum, with particularly strong ultraviolet and optical output, along with prominent spectral lines from elements such as hydrogen, magnesium, and iron.

Active Galactic Nuclei (AGN)

Quasars are the most extreme members of the active galactic nucleus (AGN) family and reside at the centers of galaxies.

The Power Source: Supermassive Black Hole

At the heart of every quasar lies a **supermassive black hole**, ranging from millions to billions of solar masses. Energy is released outside the event horizon, where infalling matter heats and radiates intensely.

How Quasars Were Found to Host Supermassive Black Holes

The conclusion that **quasars host supermassive black holes** was not reached suddenly, but emerged through a sequence of observational discoveries and physical arguments.

First, quasars were identified as extremely distant objects with **extraordinarily high luminosities,** often outshining entire galaxies. Early spectra of quasars such as **3C 273** revealed large redshifts, proving that their immense brightness originated from compact regions at cosmological distances.

Second, quasars were found to vary in brightness over **very short timescales,** sometimes days or weeks. Such rapid variability implies that the emitting region must be no larger than a region comparable to the size of the Solar System ,ruling out extended sources such as star clusters or normal galaxies.

Third, no known stellar process—nuclear fusion, supernovae, or starbursts—could account for the **enormous and sustained energy output** observed. The only physically viable mechanism was **gravitational energy released by matter falling into a deep potential well.**

Fourth, theoretical work showed that **accretion onto a compact object** could convert mass into energy with far higher efficiency than nuclear fusion. When applied to quasar luminosities, this model required a central object with a mass of **millions to billions of M⊙.**

Finally, high-resolution observations revealed **relativistic jets**, broad emission lines, and disk-like velocity structures—all features naturally explained by an accretion disk surrounding a supermassive black hole.

Taken together, these lines of evidence led to the firm conclusion that **quasars are powered by accretion onto supermassive black holes located at the centers of galaxies, establishing quasars** as the **most luminous** manifestation of active galactic nuclei.

Remarkable Quasars and Large-Scale Structures

Some quasars stand out for their extreme properties:

1. **3C 273:** A quasar is located approximately **2.4–2.5 billion light-years** from Earth in the constellation Virgo. 3C 273 is one of the closest and brightest quasars and can be observed with small telescopes. The mass of its central supermassive black hole is about **900 million M⊙.**
2. **J0529–4351:** The quasar is among the most luminous quasars ever observed. It is located roughly **12 billion light-years** from Earth in the constellation Lepus. Its central supermassive black hole has an estimated mass of about **(1.7–2.0) × 10^{10} M⊙,** making it one of the **most massive known.** The black hole is accreting matter at an extraordinary rate of approximately **one solar mass per day,** powering its extreme luminosity.
3. **3C 48:** It is among the earliest quasars to be recognized and exhibits clear signatures of galaxy merger activity in its host system. It is located about **4 billion light-years** from Earth and is associated with a powerful radio source. Its central supermassive black hole has an estimated mass of **(6–8) × 10^8 M⊙.** Observations reveal tidal distortions, intense star formation, and a disrupted galactic structure—strong evidence that quasar activity can be triggered by galaxy interactions and mergers.

Large Quasar Groups (LQGs)

In 2013, astronomers reported the discovery of enormous **Large Quasar Groups** (LQGs)—vast cosmic structures composed of dozens of quasars spanning extraordinary distances across space. One notable example is the Huge-LQG, which extends roughly **4 billion light-years** across, making it one of the largest known structures in the observable universe. Such immense groupings challenge the conventional assumption that the universe becomes homogeneous on very large scales.

Large Quasar Groups typically contain **10 to 70** or more quasars, gravitationally associated within a common large-scale filament or overdense region of the cosmic web. These structures trace regions of enhanced matter density in the early universe and provide valuable clues about large-scale cosmic evolution.

1. Quasars and Galactic Evolution

Quasars are temporary phases in a galaxy's life. Once nearby gas is exhausted, the quasar fades, leaving behind a dormant supermassive black hole. Many present-day elliptical galaxies likely passed through such an active phase.

Quasar radiation and jets can also suppress star formation by heating or expelling gas, shaping the long-term evolution of their host galaxies.

Cosmic Time Machines: Quasars are more than brilliant lights—they are windows into cosmic history. By studying them, astronomers trace the growth of black holes, the formation of galaxies, and the evolution of the universe itself. As beacons from the distant past, quasars continue to illuminate our understanding of the cosmos.

2. Blazars

A blazar is a highly energetic type of active galactic nucleus (AGN), powered by a supermassive black hole that accretes surrounding gas and dust. What makes a blazar unique is its orientation: **one of its powerful jets of charged particles is directed almost straight toward Earth.**

Discovery: Blazars were recognized in the late 1960s and 1970s when astronomers studied unusual radio sources that showed rapid brightness variations and strong polarization. Objects such as BL Lacertae and 3C 279 helped establish this new class of extremely active galaxies.

Figure 5.6: AI-generated illustration of blazar for conceptual purposes

Why Blazars Appear So Bright: Because the jet is pointed nearly toward us, their radiation is strongly amplified by an effect called relativistic beaming. This makes blazars appear exceptionally bright and energetic, even if their intrinsic power is similar to other quasars.

Blazars emit radiation across almost the entire electromagnetic spectrum—from radio waves to high-energy gamma rays.

Rapid Variability: Blazars are among the most variable objects in the universe. Their brightness can change dramatically, sometimes within hours or even minutes. These variations arise from turbulent processes within the jet, involving shock waves, magnetic fields, and high-energy particles.

They may also appear to show faster-than-light motion, but this is only an optical effect caused by material moving close to the speed of light toward the observer.

Jet Orientation — The Key Difference

Blazar: Jet points nearly toward Earth → extreme brightness and rapid variability

Quasar: Jet is angled away → emission appears steadier and less amplified

Importance: Astronomers have identified a few thousand blazars so far, though many more likely exist. Like quasars, they reside in the centers of galaxies—typically massive elliptical galaxies—and provide important insights into black hole physics, relativistic jets, and high-energy processes in the universe.

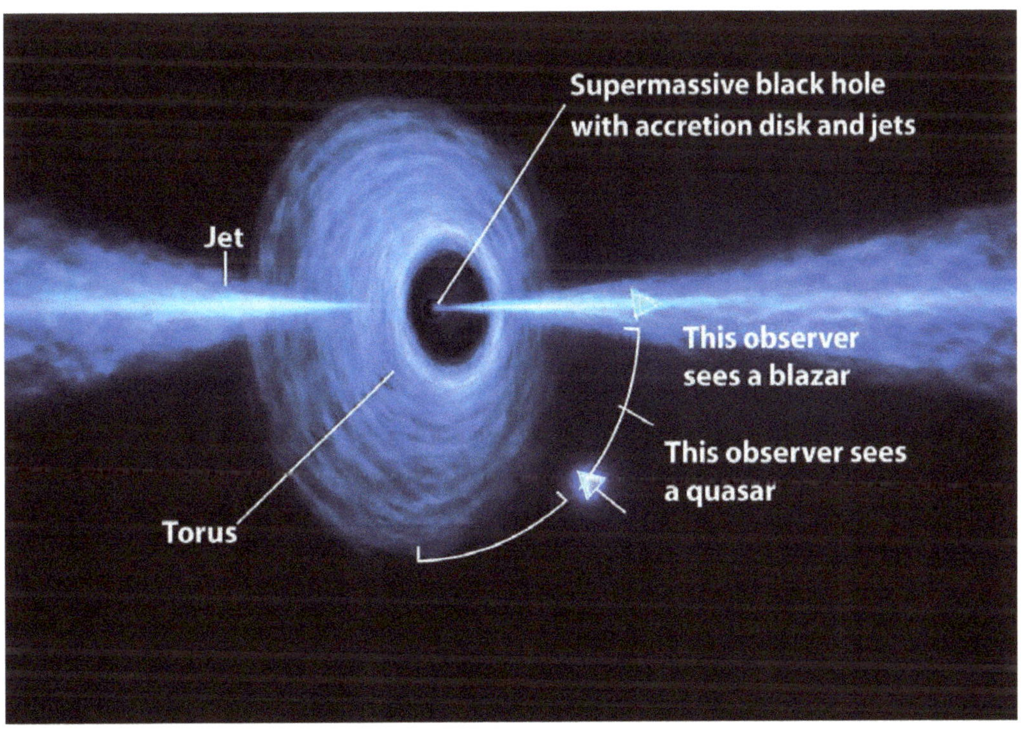

Figure 5.7: AI-generated illustration of a blazar–quasar viewing angle model

3. Rogue Black Holes

Free-Floating Wanderers of the Universe

Among the most fascinating and mysterious discoveries in modern astrophysics is the existence of **rogue black holes**, also known as **wandering** or **nomadic black holes**. These invisible cosmic drifters roam freely through space, untethered to any star, cluster, or galaxy, silently journeying across the Milky Way and beyond. Though unseen, they are among the most powerful gravitational entities in the cosmos — dark travelers shaping their surroundings through gravity alone.

Such cosmic nomad black holes are believed to form when violent cosmic events — such as **supernova explosions**, **three-body interactions**, or **galactic mergers** — eject them from their original systems. Once expelled, they travel endlessly through the galaxy, unseen but detectable by their gravitational influence.

How Are They Detected

Because rogue black holes emit no light, they are extremely difficult to detect. Their presence is typically revealed through **gravitational microlensing**, a phenomenon predicted by Einstein's theory of general relativity.

When a rogue black hole passes in front of a background star, its intense gravity bends and magnifies the starlight, causing a temporary brightening. By studying the duration and shape of this brightening, astronomers can estimate the black hole's **mass, velocity,** and **distance**.

Instruments such as the **Hubble Space Telescope (HST)**, **Gaia**, and the **Optical Gravitational Lensing Experiment (OGLE)** have been crucial in detecting these delicate gravitational signals — uncovering the invisible footprints of wandering black holes across our galaxy.

3.1 The First Strongly Favoured Rogue Stellar-Mass Black Hole: OB110462

The first strongly supported candidate for an **isolated stellar-mass black hole** was reported in 2022 after more than a decade of observations by the Hubble Space Telescope and the Optical Gravitational Lensing Experiment (OGLE) toward the Galactic Bulge. The microlensing event, designated OGLE-2011-BLG-0462 (also known as OB110462), was first detected in 2011.

The observations revealed a compact, non-luminous object located approximately **4,000–5,000 light-years** away in the direction of the constellation Sagittarius. The object was identified through gravitational **microlensing,** as its gravity temporarily magnified the light of a background star and produced a small but measurable shift in the star's apparent position.

Detailed analysis of the data, particularly from the Hubble Space Telescope, strongly favours the interpretation that the lensing object is a **solitary black hole,** detected purely through its gravitational influence rather than emitted light.

The object is estimated to have a mass of roughly **5–10 M⊙,** although some studies suggest a lower mass, and uncertainties remain. It is moving through space at a velocity of about **45 km/s.** Astronomers believe it was likely ejected during the supernova explosion that formed it, possibly due to an asymmetric core collapse that imparted a strong "**natal kick**," allowing it to escape its original stellar system.

This discovery provides some of the strongest evidence to date for the existence of **rogue stellar-mass black holes** and highlights the power of gravitational microlensing as a method for detecting otherwise invisible objects.

3.2 A Rogue Supermassive Black Hole and Tidal Disruption Event (AT2024tvd)

Figure 5.7: Rogue SMBH (TDS)

Image Credit: NASA/ESA/STScl

In 2024, astronomers reported an unusual **tidal disruption event** (TDE), designated AT2024tvd, observed in a galaxy approximately **600 million light-years** away in the constellation **Hercules**. The event was first detected as an optical transient by the Zwicky

Transient Facility and was subsequently studied using the Hubble Space Telescope, the Chandra X-ray Observatory, and other facilities.

Nature of the Event

AT2024tvd occurred when a star passed too close to a wandering supermassive black hole and was torn apart by intense tidal forces, producing a luminous flare across multiple wavelengths. Such events—known as tidal disruption events—are usually observed near the centers of galaxies, where supermassive black holes reside. However, this event was detected approximately **2,600 light-years** away from the nucleus of its host galaxy, indicating that the black hole responsible was **not located at the galactic center.**

The Black Hole: The black hole responsible for the flare is estimated to have a mass of about **~1 million M⊙.** In contrast, the central supermassive black hole of the host galaxy—located at its nucleus—has a mass of roughly **~100 million M⊙** and was **not involved in the event.**

Scientific Significance: This discovery provides strong evidence for the existence of a **wandering (off-nuclear) supermassive black hole, possibly displaced** from the galaxy's center due to past galaxy mergers or gravitational recoil following a black hole merger. It represents one of the clearest cases of an **off-nuclear tidal disruption event detected in optical surveys,** offering a new way to identify otherwise invisible wandering black holes.

3.3 A Titan Unbound: Runaway Supermassive Black Hole (Landmark Discovery of the Year 2025)

In December **2025,** astronomers confirmed one of the most extraordinary phenomena ever observed: a **runaway supermassive black hole**, hurtling through intergalactic space at an unprecedented speed. This discovery has reshaped our understanding of black hole dynamics and galaxy evolution, earning its place as one of the most remarkable astronomical milestones of the year.

The term *"runaway"* emphasizes the violent **ejection process**—a black hole forcibly expelled from its host system—whereas *"rogue"* describes its present state: **free-floating and unbound**. While the two terms are often used interchangeably, the distinction lies more in narrative emphasis than in firm scientific classification.

Discovery: The object was first identified as a compelling candidate in **2023** using observations from the **Hubble Space Telescope.** Definitive confirmation came in December 2025, following deeper infrared observations from the **James Webb Space Telescope**. The findings were led by **Pieter van Dokkum** and colleagues at **Yale University** who revealed the object's true nature and extraordinary behavior.

Figure 5.8: An artist's impression showing a runaway supermassive black hole, a galaxy-scale bow shock ahead of it, and an extended trailing wake of disturbed gas.

Image Credit: NASA Science

Location: The runaway black hole was traced to a distant and dramatic environment known as the ***"Cosmic Owl"***—a structure formed by colliding ring galaxies, located roughly **11 billion light-years** from Earth. This chaotic region likely provided the extreme gravitational conditions necessary. to eject such a massive object.

The Black Hole: Astronomers estimate the mass of this nomadic black hole to be approximately **10 million M⊙,** placing it firmly in the category of supermassive black holes. Untethered from any galaxy, it is racing through space at nearly **1000 kilometers per second**—the **highest velocity** ever recorded for a rogue black hole.

Extraordinary Features

What sets this discovery apart is not merely the black hole's speed and mass, but its astonishing interaction with its surroundings. As it plows through intergalactic gas, the black hole drives a **galaxy-scale bow shock** ahead of it, compressing matter in its path. Behind it trails an immense **wake stretching nearly 200,000 light-years,** within which gas cools, condenses, and **ignites new star formation**.

This marks the **first confirmed detection of an unbound supermassive black hole actively carrying a star-forming region with it**, roaming freely through space after a violent ejection event. Remarkably, the black hole now lies about **230,000 light-years** away from its former host galaxy, more than **twice** the width of the Milky Way galaxy making it the most distant displaced SMBH ever observed.

This confirmed a runaway supermassive black hole, now named as **RBH-1.** Researchers believe that the event which gave **RBH-1** its kick across spacetime, was most likely the gravitational recoil from supermassive black hole merge predicated by theory of relativity but rarely found. Its discovery has electrified the astronomical community. The runaway supermassive black hole stands as a powerful reminder that even the universe's most massive anchors can be torn.

Cosmic Implications: Astronomers believe that thousands of such runaway supermassive black holes may wander the universe, largely unseen. Future wide-field surveys—particularly with the upcoming — **Nancy Grace Roman Space Telescope** are expected to uncover many more of these cosmic exiles.

3.4 3C 186: Rogue Supermassive Black Hole Candidate

In 2017, astronomers reported an extraordinary discovery: a supermassive black hole that appears to be displaced from the center of its host galaxy.

This object is associated with the quasar **3C 186** approximately **8 billion light-years** away in the constellation **Ursa Major.** Observations with the Hubble Space Telescope revealed that the quasar's supermassive black hole is offset by about **35,000 light-years** from the luminous core of its host galaxy.

Black Hole: The black hole has an estimated mass of roughly **1 billion M⊙.** Despite being displaced from the galactic center, it remains actively accreting matter, as evidenced by its powerful quasar emission. Its sustained activity demonstrates that a supermassive black hole can continue to draw in gas and radiate enormous amounts of energy even while traveling through its host galaxy.

Such a large displacement is highly unusual, as supermassive black holes are generally found at the centers of galaxies, where they reside within the gravitational potential well of the stellar bulge.

The leading explanation is that the SMBH was **ejected from the galactic nucleus** following a **merger** with another **black hole**, with gravitational waves imparting a "kick" that sent it drifting through space. This process, known as a **gravitational recoil**, is predicted by general relativity but is rarely observed.

Scientific Significance: 3C 186 represents one of the strong candidates for a rogue supermassive black **hole**. While the **offset** is confirmed, the interpretation of the black hole as **"rogue"** — unbound and displaced by gravitational-wave recoil — is strongly supported but not absolutely proven. Continued observations are needed to fully understand its motion and history.

This discovery has important implications:

1. It provides observational support for black hole ejection mechanisms predicted by simulations of SMBH mergers.
2. It challenges the traditional view that all supermassive black holes are confined to galactic centers.
3. It demonstrates that rogue SMBHs can remain active and highly luminous, reshaping their cosmic environment even far from the galactic nucleus.

Drifting silently through interstellar and intergalactic space, they are the true nomads of the cosmos — unseen, yet leaving gravitational footprints that remind us that even in the deepest void, the universe is never still.

4. White Hole

A white hole is a theoretical concept in physics, often described as the reverse of a black hole. While black holes are regions of space where nothing—not even light—can escape, white holes are hypothetical regions from which **matter, energy, and information can escape**, but **nothing can enter**.

White holes emerge as mathematical solutions to **Einstein's equations of general relativity**, They are predicted to possess an event horizon analogous to that of a black hole, though in this case it functions as a past event horizon, permitting outward flow while forbidding inward passage.

Figure: 5.9: Conceptual illustration of a white hole- a hypothetical time-reversed counterpart of a black hole, expelling matter and energy instead of absorbing it.

Despite their presence in theoretical physics, **no observational evidence** for the existence of white holes has ever been found. They remain purely speculative. Some theories propose that white holes might be connected to black holes via **wormholes**, forming a tunnel through spacetime—a concept known as an **Einstein-Rosen bridge**. However, this idea also remains highly theoretical.

Other proposals suggest that white holes could be **inherently unstable and short-lived**, which would make them extremely difficult to detect, even if they do exist. Their instability could cause them to collapse or vanish almost as soon as they form. Therefore, while white holes remain an intriguing prediction of theoretical physics, no direct observational evidence has yet confirmed their existence.

Extreme Limits Black Holes

Even the most powerful black holes are governed by fundamental physical limits that restrict how fast they can grow and how brightly they can shine.

The Eddington Limit

The **Eddington limit**, named after the English astronomer **Sir Arthur Eddington**, defines the maximum luminosity an astrophysical object—such as a star or black hole—can achieve when the **outward force of radiation pressure exactly balances the inward pull of gravity**. If an object exceeds this limit, radiation pressure can repel infalling material, effectively **regulating** or **halting** further accretion.

Eddington Accretion Rate: It is the maximum accretion rate at which matter can fall on to a black hole.

Eddington Accretion Rate Formula (approximate)

$$\dot{M}_{Edd} \approx 1.4 \times 10^{18} (M/M_{\odot})\ gs^{-1}$$

Where

$\dot{M}_{Edd}$ —- Eddington accretion rate

M ----------- mass of a black hole

$M_{\odot}$ ----------mass of Sun

gs^{-1} ---------- gram per second

This expression **assumes a radiative efficiency of about 10%,** which is standard for accretion onto black holes.

(Radiative efficiency is a measure of how effectively a black hole converts falling matter into energy.)

Maximum Luminosity: The highest energy output a celestial body can sustain without expelling mass due to radiation pressure.

Significance: The Eddington limit serves as a theoretical cap on black hole growth under normal conditions, so violations of this limit suggest that **non-standard accretion mechanisms** may be at play—such as anisotropic radiation, photon trapping, or dense inflow geometries.

Super-Eddington Accretion

A remarkable black hole known as **LID-568** has been discovered accreting matter at an extraordinary rate. Observed by the James Webb Space Telescope (JWST) at a distance of about **12.1 billion light-years**, it is seen less than **one billion years** after the Big Bang, offering insight into the early growth of supermassive black holes.

With a mass of roughly *7* **million solar masses**, LID-568 appears to accrete at up to **ten times the classical Eddington limit**—the point where radiation pressure balances gravity. This suggests that under extreme conditions, black hole growth can exceed traditional theoretical limits. The mechanisms behind such super-Eddington accretion remain an active area of research.

Cosmological Coupling

Black Holes and the Expanding Universe

Cosmological coupling refers to a proposed phenomenon in which black hole mass slowly increases as the Universe expands. Unlike traditional growth mechanisms—such as swallowing gas, dust, or merging with other black holes—this hypothesis suggests that black holes could gain mass passively, as a consequence of their connection to the geometry of the cosmos.

Theoretical Foundation: The speculative idea of cosmological coupling is inspired by **Albert Einstein's General Theory of Relativity,** which describes gravity as the curvature of spacetime and links matter, energy, and geometry into a single framework.

Some researchers have suggested that black hole mass growth might arise naturally from solutions of Einstein's equations when cosmological boundary conditions are taken into account. In this view, black holes would not be entirely independent of the Universe's global structure, and their properties might evolve slowly alongside cosmic expansion.

Observational Motivation

Interest in cosmological coupling increased after studies of supermassive black holes in large elliptical galaxies. These galaxies were chosen because they are typically inactive, showing little gas, weak or absent accretion disks, and no strong signs of recent mergers.

Surprisingly, observations suggest that the masses of some supermassive black holes may have increased by factors of about **7** to **20** over the last **9 billion years**, even in systems where traditional growth mechanisms appear minimal. This apparent mass increase seems to correlate with the expansion history of the Universe.

This observation has been interpreted by some researchers as tentative evidence that black hole mass might evolve in step with cosmic expansion.

However, alternative explanations remain possible, including hidden accretion, ancient mergers, and uncertainties in mass measurements. Therefore, these observations are intriguing but not yet conclusive.

Connection with Dark Energy

Some scientists have proposed a possible link between black holes and dark energy, the mysterious force causing the universe to expand faster.

In this idea, black holes might interact with the energy of empty space (vacuum energy), possibly gaining mass from it. If this is true, black holes may not be just passive objects—they could play a small role in the expansion of the universe.

This is still a speculative idea, but if confirmed, it could change our understanding of both black holes and the universe.

Implications and Future Research

If cosmological coupling is confirmed, it would redefine black holes as more than just endpoints of stellar evolution or galactic centers. They would become dynamic participants in the evolution of the Universe.

Such a discovery would:

1. Provide a possible new perspective on dark energy and its origin.
2. Revise models of black hole and galaxy evolution, especially in quiet, isolated systems.
3. Deepen the connection between gravity, cosmology, and quantum vacuum physics.

As observations improve, especially through gravitational wave astronomy, high-precision black hole mass measurements, and advanced simulations, cosmological coupling may be tested more rigorously. It could either be ruled out or emerge as a fundamental feature of nature.

At present, no universally accepted physical theory predicts a direct increase in black hole mass or spin caused purely by the expansion of space. Cosmological coupling therefore remains an exciting and thought-provoking hypothesis at the frontier of modern astrophysics, not yet a confirmed part of black hole physics.

Concluding Remarks

From the intimate dance of the V404 Cygni triple system to the extreme environment of the Sagittarius A*–D9 system, this chapter has explored how black holes rarely exist in isolation. Quasars and blazars reveal how feeding black holes can light up the universe across billions of years.

Rogue black holes remind us that some of these objects wander the galaxy unseen, while concepts such as the Eddington limit and cosmological coupling show how fundamental laws govern their growth and influence. Even the speculative idea of white holes pushes us to question what lies beyond established theory.

Together, these objects and ideas form a connected cosmic narrative—one in which black holes act not only as endpoints of stellar evolution, but as powerful agents shaping galaxies, radiation, and the large-scale structure of the universe itself.

Chapter 6
Physical Characteristics and Extreme Gravitational Effects of Black Holes

Introduction

Black holes serve as natural laboratories for studying gravity, matter, and spacetime under conditions that cannot be reproduced anywhere else in the cosmos. Their physical properties are extraordinary—immense **density**, rapid **spin**, extreme **low temperature**, and powerful **magnetic fields** define their behavior and influence their surroundings.

The gravity around a black hole is so intense that it warps space and time, producing remarkable effects such as **gravitational redshift**, **gravitational lensing**, and **time dilation**. Close to the event horizon, the gravitational pull becomes so strong that even atoms and light are stretched and distorted, a process known as **spaghettification**.

By studying these physical and gravitational characteristics, astronomers gain vital information about the **nature, structure, and evolution** of black holes. Each feature—from density to extreme gravity—reveals how these cosmic enigmas shape the universe around them and test the very foundations of Einstein's theory of relativity.

Physical Characteristics

1. Density of Black Holes

The concept of density in black holes presents a unique challenge in theoretical astrophysics. Unlike ordinary matter, a black hole does not have a well-defined internal structure due to the attack of its core—a point at which classical general relativity predicts infinite density and zero volume. Because of this, the central density of a black hole is not physically meaningful in the traditional sense.

However, astrophysicists can calculate an **average density** by considering the volume enclosed within the **event horizon**—the boundary beyond which no information or matter can escape. This average density provides useful comparative insight, especially when analyzing black holes of different masses.

Mathematical Derivation: Average Density of a Black Hole

Let **M** be the mass of a non - rotating black hole(Schwarzschild black hole)and let **R** be the Schwarzschild radius (the radius of the event horizon), which is given by:

$$\mathbf{R = 2GM/c^2}$$

Assuming spherical symmetry,

Average Density(ρ)= Mass/ Volume

$$\rho = \mathbf{M/4/3\ \pi\ R^3}$$

Substituting $\mathbf{R = 2GM/c^2}$

$$\rho = \mathbf{3c^6/32\pi G^3M^2}$$

$$= \mathbf{K/M^2}$$

(Where $K = 3c^6/32\pi G^3$)

$$\rho = \mathbf{K/M^2}$$

$$\rho \propto \mathbf{1/M^2}$$

The average density of a black hole is inversely proportional to the square of its mass.

This counterintuitive result reveals that:

Stellar-mass black holes possess extremely **high average densities**, while **intermediate-mass black holes** can have **densities** comparable to that of **water**.

Supermassive black holes at galactic centers may exhibit **average densities lower** than that of **air,** despite their immense gravitational influence.

While classical general relativity predicts infinite density at the central singularity, the average density within the event horizon decreases rapidly as black hole mass increases.

This surprising result highlights a key distinction in black hole physics: strong gravity is not necessarily associated with high density. Instead, gravitational intensity arises primarily from the curvature of spacetime itself, rather than from simple mass concentration. This challenges conventional notions of compactness and deepens our understanding of gravity in extreme regimes.

For rotating (Kerr) black holes, an exact density is not well defined because the horizon is not spherical. However, an effective average density can be estimated using the spin-dependent event-horizon radius, and it still decreases with increasing black hole mass.

2. Spin of the Black Hole

Black holes are not understood as material objects with internal mechanical structure. In general relativity, they are fully described by **the geometry of spacetime** — specifically by their mass, angular momentum, and charge. When we say that a black hole rotates, we mean that its angular momentum is encoded in the structure of spacetime itself, causing spacetime to twist. Thus, black holes are fundamentally **geometric objects** rather than conventional physical bodies made of matter.

What Does It Mean for a Black Hole to Rotate

When we say that a black hole "rotates," we do not mean that it spins like a solid ball, because a black hole has no physical surface or rigid body. Instead, a rotating black hole possesses angular momentum, and in Einstein's theory of general relativity this angular momentum causes the very fabric of spacetime around it to twist. The effect is known as frame dragging: space and time themselves are pulled around in the direction of the black hole's spin. Near the black hole, this twisting becomes so strong that nothing—not even light—can remain stationary, and everything is compelled to move in the direction of rotation.

The event horizon, which is not a material surface but a boundary defined by the paths of light, also shares this rotational property because the spacetime at that boundary is rotating. When we say the event horizon has angular velocity, we mean that the spacetime geometry at the horizon is rotating-not that a physical surface is spinning.

Thus, when scientists speak of a spinning black hole, they mean that the **geometry of spacetime itself carries angular momentum** and exhibits rotation, not that a solid object is physically turning.

The **spin** or **angular momentum** of a black hole is a fundamental and crucial characteristic that significantly influences its environment, affects the formation of jets, governs its evolution, and plays a key role in ongoing research on black hole behavior and the dynamics of their accretion disks.

Why Do Black Holes Spin

Black holes are typically formed from the gravitational collapse of massive stars or from the merger of compact objects like neutron stars or smaller black holes. Since all stars and colliding objects possess intrinsic rotation, the black holes they form are naturally expected to spin.

When a star collapses into a black hole, it spins faster due to the conservation of angular momentum. The initial rotational energy is compressed into a much smaller volume, increasing the rotational speed significantly in the absence of external torque—similar to how a figure skater spins faster when pulling in their arms.

Relationship Between Mass and Spin

Accretion and Spin

A black hole's spin can provide important clues about its formation history and growth. Black holes grow by accreting matter, and as this matter spirals inward, it carries angular momentum. This causes the black hole to spin faster or lower.

Low-Mass SMBHs: In the case of lower-mass supermassive black holes (SMBHs), higher spin values often suggest that they grew primarily through **coherent accretion** of matter over time—where the incoming matter consistently falls in along a single direction.

High-Mass SMBHs: For more massive SMBHs, high spin values are less common. This implies that their growth may have occurred through multiple chaotic mergers, where the spin directions of the incoming objects vary, often canceling out and leading to lower net spin values.

Spin from Mergers: When two black holes merge, the spin of the resulting black hole depends on the masses, individual spins, and the orientation (angle) of their spins during the merger. Such events can either increase or decrease the final spin.

If the merging black holes have aligned spins, the resulting black hole can spin faster.

If their spins are misaligned or in opposite directions, the final spin may decrease.

During mergers or accretion, the total amount of rotation in a system is conserved. This rotation may come from two merging black holes or from matter swirling in an accretion disk around a black hole. How this rotation is shared determines how fast the final black hole spins. More massive black holes change their spin more slowly because their greater mass resists change.

In some cases, a black hole can spin opposite to the direction of the material spiraling into it. This is known as **retrograde spin**. Retrograde configurations affect how close matter can orbit the black hole and influence the efficiency of jet production and accretion.

In contrast, when a black hole's spin is **in the same direction** as the rotation of the accreting material, it is called **prograde spin**. In a prograde configuration, the angular momentum of the black hole and the accretion disk are aligned.

This alignment allows matter to orbit much closer to the event horizon before crossing it, compared to a retrograde system. As a result, the **innermost stable circular orbit (ISCO)** lies nearer to the black hole, enabling the accreting material to release more gravitational energy as radiation. Prograde spin is often associated with **higher accretion efficiency** and can produce **powerful relativistic jets**, since the rotational energy of the black hole can be more effectively extracted through processes such as the **Blandford–Znajek mechanism**.

Jets and Spin (AGN Jets)

The spin of a black hole is believed to play a critical role in launching **jets** from active galactic nuclei (AGN). These jets, made of energetic particles and radiation, are typically ejected along the spin axis of the black hole.

Through processes such as the **Blandford–Znajek mechanism**, a spinning black hole can transfer some of its rotational energy to power these jets, gradually slowing its spin over time.

Mathematics of Spin

We introduced black hole spin earlier when discussing rotating black holes. Here we define it precisely and show how scientists quantify it.

The spin of a black hole is quantified by a **dimensionless spin parameter**, denoted by **"a*"**

$$a^* = Jc/GM^2$$

where

J - angular momentum of the black hole

M - mass of the black hole

G,c - fundamental constant of nature

A black hole with **1 or -1** is said to be spinning at the **maximum possible rate**.

Such a spin causes the event horizon to shrink to its minimum radius, and matter can orbit closer—moving at speeds near that of light.

Spin Limit: While a black hole can spin increasingly faster by accreting matter, there exists a theoretical upper limit to its spin. Based on the **Kerr metric**, the **maximum spin** is limited to **1 or -1.** If this limit is exceeded, it would result in a **naked singularity**—a point of infinite density exposed to the universe—which is **not predicted** by current theories of general relativity.

In reality, radiation pressure and magnetic effects from the accretion disk likely prevent a black hole from ever reaching.

Observational Evidence: Observations of X-ray binaries and supermassive black holes suggest that black holes can attain spin values close to, but not exceeding, the theoretical maximum. Such measurements support the predictions of Einstein's theory and provide insights into the black hole's past.

How is Spin Measured

Astronomers use several techniques to estimate the spin of a black hole:

1. **X-ray Spectroscopy:** This technique studies the detailed "fingerprints" in X-ray light coming from the hot gas very close to a black hole. One of the most important features is the iron emission line. Near a black hole, this line becomes stretched and distorted because of intense gravity and rapid motion.

   The amount of distortion tells scientists how close the glowing gas can orbit the black hole. Since faster-spinning black holes allow matter to orbit closer in, the shape of these X-ray features reveals how rapidly the black hole is rotating.

2. **Continuum Fitting:** Continuum fitting compares the observed X-ray spectrum of the hot accretion disk with theoretical models to find how fast the black hole is rotating.

   By carefully measuring the overall shape and temperature of this X-ray light, scientists can understand how the disk is structured and how close it extends toward the central object.

   From this "glow" of the disk, astronomers can estimate important properties of the black hole, such as its spin and the size of the region where the disk ends.

3. **Jet modeling:** Analyzing the power and orientation of jets in AGN to estimate the spin based on energy extraction models.

   These observations reveal how matter and light behave near the innermost stable circular orbit (ISCO), which depends on the spin.

4. **Dragging:** Frame dragging is a fundamental relativistic effect produced by the rotation of a black hole and is therefore an intrinsic physical characteristic.

   When a massive object like a planet or a black hole spins, it doesn't just rotate on its own axis — it actually drags the fabric of space and time around with it. This strange effect is called frame dragging.

   You can imagine space-time as a thick, invisible syrup. If you place a spinning ball into that syrup, the fluid nearby will start to swirl around it. Similarly, a rotating black hole makes the surrounding space-time twist in the direction of its spin.

Figure 6.1: AI-generated Illustration of Frame Dragging for conceptual purposes

Because of this, anything — light, matter, or even another smaller black hole — that comes too close is forced to move along with the spinning space-time. This region, where nothing can remain still, is called the ergosphere. Inside the ergosphere, even light is dragged around the black hole's rotation, although it can still escape if it moves in the right direction.

The angular momentum or spin of a black hole can be measured from a distance through the phenomenon of **frame dragging,** caused by the black hole's **gravitomagnetic field.** This effect produces a tiny **relativistic precession** in the orbital plane of a nearby object, resulting from the rotation of the massive central body. Frame dragging is one of the key predictions of Einstein's theory of general relativity in the weak-field and slow-motion limit, and it is also known as **the Lense–Thirring effect.**

Significance of Spin

Understanding black hole spin is vital for:

1. Tracing their **formation** and **growth**.
2. Studying their influence on **galactic evolution** and **feedback processes**.
3. Exploring high-energy phenomena like **jets, quasars, and gravitational waves.**

Spin and the Information Paradox: Black hole spin may also have implications for the **black hole information paradox**—the puzzle of whether information about matter falling into a black hole is truly lost or somehow preserved. Spin affects how black holes radiate, interact with their environment, and potentially encode information in quantum processes. The spin of a black influences how it interacts with its surroundings,shaping accretion disks.

Why is Charge Usually Omitted

Although black holes can theoretically carry an electric charge,no charged black hole has ever been observed in nature. In realistic astrophysical environments, any charge would be neutralized by surrounding plasma and infalling matter. Therefore, charge is not considered among the physical characteristics of black holes which are instead defined by their mass, spin, temperature, density and entropy.

3. Temperature of Black Holes

Black holes, despite their reputation for extreme gravitational intensity, are among the coldest known objects in the observable universe. Their physical temperature, governed by **Hawking radiation**, is remarkably close to **absolute zero**. This temperature is **inversely proportional to the mass** of the black hole—**more massive black holes are colder**, while **smaller black holes are hotter**.

For example

A stellar-mass black hole ~**10 M⊙** has a temperature in the range of **10^{-8} K.**

A supermassive black hole ~**10^9 M⊙** have a temperature of only **1.4×10^{-14} K**.

By contrast, hypothetical micro black holes—possibly formed in the early universe—would be extremely hot and could emit intense radiation. However, no such objects have ever been detected, and they remain purely theoretical.

Temperature of Accretion Disk

The temperature of the accretion disk of a supermassive black hole can reach **10 million** degree Celsius and a stellar-mass black hole's accretion disk can reach ever higher.

X- ray binaries: In these systems the temperature of the inner disk can be of the order **10^7 K.**

Thus, while the environment around a black hole can be incredibly hot, the black hole itself remains extremely cold.

Hawking Radiation and Temperature Formula

Physicist Stephen Hawking, in 1974, theorized that black holes are not completely black but instead emit a faint radiation due to **quantum effects near the event horizon**—a phenomenon now known as **Hawking radiation**. This emission gives rise to a **black hole temperature**, described by the formula. For non-rotating , uncharged black hole, the Hawking temperature is given by formula

$$T_H = c^3\hbar/8\pi k_B GM$$

$$T_H = hc^3/16\pi^2 k_B GM$$

$T_H = K/M$

(Where $K = hc^3 / 16\pi^2 k_B G$)

$T_H \propto 1/M$

Where

T_H - the Hawking temperature

(in kelvin K)

M - mass of the black hole.

k_B - Boltzmann constant

$\bar{h}$ - reduced Planck's constant

G, c, h - fundamental physical constants

This shows that the **temperature of black hole is inversely proportional to its mass.**

Larger black holes radiate less and thus are **colder.**

Smaller black holes radiate more and lose mass over time, leading to rapid evaporation and a potential **final burst of high-energy radiation**.

Hawking temperature and Event horizon radius

$R_h = 2GM/c^2$

$GM = R_h c^2/2$

Substituting in equation

$T_H = hc^3/16\pi^2 k_B GM$

$T_H = hc/8\pi^2 k_B R_h$

$T_H \propto 1/R_h$

Thus Hawking temperature is inversely proportional to the event horizon radius.

This means:

Smaller black hole → smaller event horizon → **higher temperature**

Larger black hole → larger event horizon → **lower temperature**

However, no direct observational evidence of Hawking radiation has yet been detected. In particular, the evaporation signatures expected from primordial black holes have not been observed. Thus, Hawking radiation remains a powerful theoretical prediction, awaiting experimental confirmation.

Why is Hawking Radiation Hard to Detect

1. **Extremely Low Temperature:** A black hole's temperature is **inversely proportional to its mass.** For stellar-mass black holes, the temperature is only a tiny fraction above absolute zero- far colder than the cosmic microwave background glow that fills the universe. This means the faint Hawking radiation is drowned out by the stronger background radiation.

2. **Very Weak Signal:** Hawking radiation is incredibly faint compared to the other emission around the black holes such as X-rays from hot gas in the accretion disk or jets of plasma. These stronger signals completely overshadow the tiny quantum radiation, making it impossible to separate.

3. **Large Black Hole Lives Too Long:** A large black hole evaporates extremely slowly. The lifetime of a black hole is much longer than the age of the universe- so the evaporation process is invisible to us.

4. **Primordial Black Holes are Hypothetical:** The tiny black holes would evaporate rapidly and emit detectable radiation. However no such primordial black holes have been found, so we have no direct source to study.

5. **Cosmic Interference:** Even if Hawking radiation exists, it would be swamped by cosmic noise.

6. **Low Cross-sectional Area:** Hawking radiation has extremely low cross- sectional area. Even if all maintained obstacles are removed, still we have no telescope to detect such low cross-sectional area radiation.

Hence detecting such faint radiation in this "cosmic fog" is like trying to hear a whisper in the middle of a rocket concert.

Hawking radiation is theoretically unavoidable but practically invisible for the kinds of black holes we know exist today. Only very small black holes may or may not exist that would radiate strongly enough for us to detect.

4. Entropy of a Black Hole

In everyday thermodynamics, entropy is a measure of disorder. A key law of physics says that the entropy of the universe can never decrease — it always increases with time.

When black holes were first studied by scientists like **Karl Schwarzschild** and **John Wheeler**, they were thought to have **zero entropy**. That meant black holes were treated as objects of **perfect order**, with no thermodynamic behavior at all. But this view created a serious problem: if an object with its own entropy fell into a black hole, the total entropy of the universe would go down — violating the laws of thermodynamics.

To solve this paradox, in **1972 Jacob Bekenstein** proposed that black holes must themselves have entropy. He argued that since black holes grow by pulling in matter and energy, their entropy should increase as well.

At first, **Stephen Hawking** disagreed. But in 1974, after his own discovery of **Hawking radiation**, he combined his work with Bekenstein's ideas. Together, they showed that black holes do follow thermodynamic laws:

The sum of the entropy outside the black hole plus the entropy of the black hole itself can never decrease.

Bekenstein–Hawking Entropy Formula

$$\mathbf{S = k_B\, Ac^3/4G\hbar}$$

$$\mathbf{S = KA}$$

Where

$$\mathbf{K = k_B\, c^3/4G\hbar}$$

$$\mathbf{S \propto A}$$

Where

S – entropy of the black hole

A – surface area of event horizon

$\mathbf{k_B}$ –Boltzmann's constant

G,c – fundamental physical constant

Ħ – reduced Planck constant

Bekenstein showed that **the entropy of a black hole is directly proportional to the area of its event horizon**, not its volume. In other words, the more surface area a black hole has, the more disorder (entropy) it contains.

The Bekenstein–Hawking entropy formula applies to all stationary black holes, including non-rotating (Schwarzschild), rotating (Kerr), charged (Reissner–Nordström), and rotating–charged (Kerr–Newman) black holes. While the mathematical expression for the event-horizon area differs in each case, the entropy always remains proportional to the horizon area. This principle has been experimentally confirmed through observations of merging black holes, which show that the event-horizon area of the final black hole is always greater than the sum of the horizon areas of the two original black holes. This implies that the entropy of black holes always increases, providing direct confirmation that black holes are genuine thermodynamic systems and truly possess entropy.

Bekenstein-Hawking's discovery established black holes as **thermodynamic systems**, possessing **entropy, temperature**, and the ability to **emit radiation**, linking black hole physics with **quantum mechanics** and **thermodynamics.**

5. Magnetic Fields Around Black Holes

Although black holes themselves cannot possess magnetic fields in the traditional sense—since they are regions of spacetime defined only by mass, and angular momentum—the environments around them are often rich in powerful magnetic activity. These magnetic fields are not intrinsic to the event horizon or the singularity, but are instead generated by the plasma and charged particles in the black hole's vicinity, particularly in the accretion disk and jet structures.

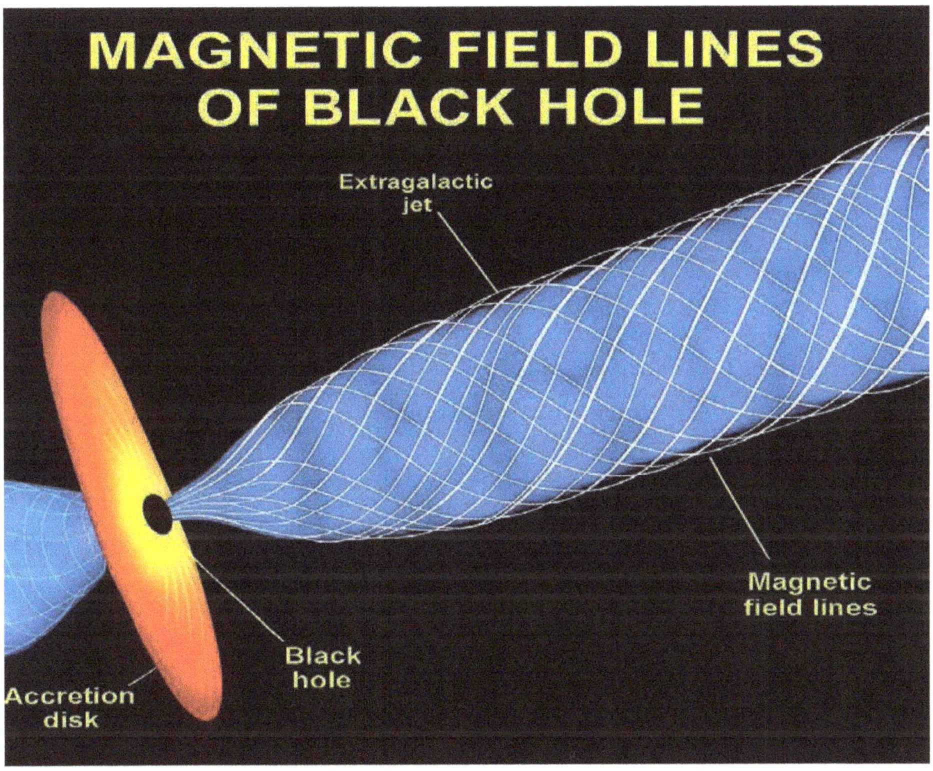

Figure 6.2: AI-generated illustration of magnetic field lines of black hole for conceptual purposes

As matter falls toward a black hole, it does not plunge directly inward. Instead, due to the conservation of angular momentum, infalling material forms a rotating structure known as an accretion disk. This disk is composed of ionized gas—plasma—rich in charged particles, which orbit the black hole at increasingly high speeds as they spiral inward and generate magnetic fields through the magnetohydrodynamics process.

Through a mechanism known as the magnetorotational instability (MRI), differential rotation within the disk amplifies weak seed magnetic fields, generating a powerful and often turbulent magnetic field throughout the disk.

Magnetic Fields as Drivers of Powerful Jets

The magnetic field lines in the accretion disk become twisted and sheared due to the intense gravitational and rotational forces near the black hole. These magnetic fields can exert pressure that counteracts the inward pull of gravity, temporarily halting accretion in some regions and redirecting energy outward. This dynamics gives rise to some of the most energetic phenomena observed in the universe, relativistic jets. These jets, which can extend well beyond the host galaxy, are believed to be launched and collimated by magnetic field lines that are anchored in the inner accretion disk and twisted by the black hole's spin—a mechanism sometimes referred to as the **Blandford–Znajek process**.

Furthermore, recent simulations and theoretical models suggest that magnetic flux can accumulate near the event horizon, creating what is known as a **Magnetically Arrested Disk (MAD)**. In this state, the magnetic pressure becomes so dominant that it **chokes off accretion temporarily, resulting in extremely luminous jet activity.** Observations from instruments such as the Event Horizon Telescope (EHT) support the existence of such strong magnetic structures. The EHT's groundbreaking image of the supermassive black hole in the galaxy M87*, for instance, revealed patterns of polarized light consistent with strong, organized magnetic fields in the inner accretion flow.

Jets, Winds, and Star Formation

The jets and winds driven by magnetic fields can heat and displace interstellar gas, suppressing or triggering star formation in a process known as **feedback**.

Stellar Collapse and Magnetic Field Formation

Interestingly, the origin of the magnetic fields observed near black holes may trace back to the progenitor stars that formed them. As a massive star collapses into a black hole during a supernova or other catastrophic event, any existing magnetic field is compressed and intensified. While the black hole itself cannot retain this field internally, the surrounding plasma can carry and amplify it through rotational and turbulent processes in the accretion disk.

This mechanism is thought to regulate the growth of galaxies over cosmic time. In fact, the correlation observed between the mass of a galaxy's central black hole and the properties of its bulge suggests that black holes and their magnetic fields play a fundamental role in galactic evolution.

Magnetic Field Strength in Black Hole Systems

The magnetic field around a black hole depends mainly on its mass, spin, and accretion rate. It is shaped and amplified within the accretion disk through processes such as turbulence and dynamo effects, and is typically strongest near the inner regions of the disk.

Dynamo Effect

The dynamo effect in the accretion disk converts the motion of ionized gas into magnetic energy, strengthening magnetic fields that help launch powerful jets.

Black holes, through their accretion disks and relativistic jets, are associated with some of the strongest magnetic fields in the universe.

Magnetic Fields Around Black Holes

Typical magnetic field strengths near the inner regions (close to the ISCO) vary with black hole mass:

Stellar- mass black hole: (~5–30 M⊙):

Magnetic field strength ≈ 10^7–10^8 G

Intermediate-mass black hole: (~10^2–10^5 M⊙):

Magnetic field strength ≈ 10^4–10^6 G

Supermassive black holes: (~10^6–10^{10} M⊙)

Magnetic field strength ≈ 10^2–10^4 G

Magnetic Field Strength and Black Hole Mass:

Magnetic field strength depends on how concentrated energy is near the black hole. Smaller black holes have compact accretion disks, where energy and plasma are tightly packed, producing strong magnetic fields. In contrast, supermassive black holes have much larger disks, so the energy is spread over a wider region, leading to comparatively weaker magnetic fields.

Approximate magnetic field strength near a black hole (Blandford–Znajek scaling)

$B \approx 10^8(\dot{M}/\dot{M}_{Edd})^{1/2}(M_\odot/M)^{1/2}$ G

Where

B------------- magnetic field near EH

Ṁ———------- mass accrction ratc

Ṁ_{Edd} — Eddington accretion rate

M⊙———----- mass of the Sun

M——--------- Mass of the black hole

This is an order-of-magnitude scaling relation, not an exact law.

It assumes: a rotating (Kerr) black hole,an accretion disk near the Eddington rate, magnetic fields supplied by the disk.

This expression gives an approximate value for the magnetic field strength threading the region near a spinning black hole. The field becomes stronger when the accretion rate is higher and weaker for more massive black holes. This magnetic field plays a central role in the Blandford–Znajek mechanism, which explains how energy and powerful jets can be extracted from a rotating black hole.

Blandford–Znajek Jet Power

(Scaling form)

$\mathbf{P_J \propto B^2\, r_g^{\,2}\, c\, a^{*2}}$

$\mathbf{P_J = K\, B^2\, r_g^{\,2}\, c\, a^{*2}}$

Where

P_j — power of relativistic jet,

B —- magnetic field near event horizon

c —- velocity of light

r_g = GM/c^2 (gravitational radius)

a* — dimensionless spin parameter

K is a dimensionless factor that depends on magnetic field geometry and plasma conditions.

Typical values from simulations are as **K= 0.05-0.1**

This formula shows how jet power increases strongly with the magnetic field strength, black hole's mass and especially on the spin. Faster spinning black holes can launch extremely powerful relativistic jets that extend across thousands of light-years.

In summary, while black holes themselves do not harbor magnetic fields, the space around them—dominated by accreting plasma and intense gravitational energy—is a crucible for magnetic field generation and amplification. These magnetic fields are not only central to the dynamics of the accretion process but also drive the powerful jets that allow black holes to shape their host galaxies across vast cosmic distances.

Effects of Extreme Gravity

The gravity of a black hole is not just a force—it is the ultimate expression of curved spacetime. Near the event horizon, gravity becomes so intense that it bends light, stretches time, and pulls matter apart. All the physical phenomena surrounding black holes—from extreme temperatures to the distortion of time—arise from this single, defining feature "**Gravity of Black Hole**". This immense distortion shapes everything nearby bending of light, stretching of time and defining the remarkable phenomenon that follows.

1. Gravitational Redshift

Gravitational redshift is a phenomenon predicted by Einstein's theory of general relativity, in which light or other electromagnetic radiation loses energy as it escapes from a gravitational field. This loss of energy causes the light's **frequency to decrease** and its **wavelength to increase**, making it shift toward the red end of the spectrum. The effect becomes more significant near extremely dense and massive objects such as neutron stars or black holes.

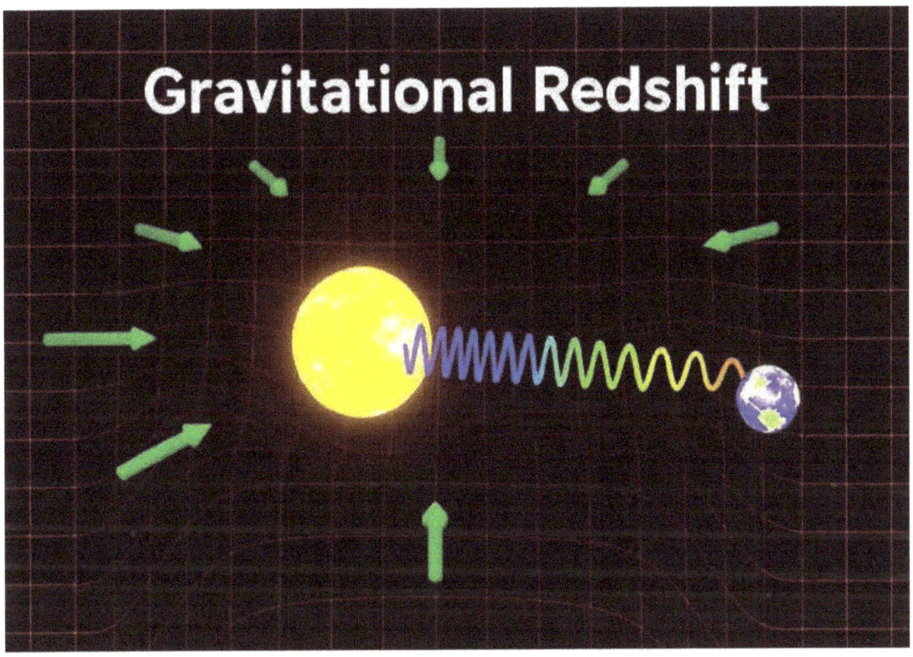

Figure 6.3: AI-generated illustration of Gravitational Red shift for explanatory purposes

The image shows light closer to the quasar appears bluer, while the light reaching an observer on Earth is redshifted due to the loss of energy during its journey.

In general relativity, gravity arises from the curvature of spacetime caused by mass and energy. This curvature alters the path and energy of light traveling through it. Near a black hole, a photon will have to expend energy to escape the gravitational pull. As it approaches the **event horizon**, the required energy becomes so large that the photon can no longer escape, and its frequency is effectively redshifted toward zero.

Real-World Evidence

This observational confirmation comes from the star **S2**, which orbits the supermassive black hole **Sagittarius A*** at the center of our galaxy. During its closest approach, astronomers detected a gravitational redshift in the light from S2 that matched the predictions of **Einstein's general relativity**. This provided one of the strongest pieces of evidence for the effect in a real astrophysical environment.

Gravitational Redshift and Black Holes

Gravitational redshift is an essential tool in the study of black holes because it directly reflects the strength of their gravitational field. When light is emitted from matter near a black hole—such as from the **accretion disk**—it becomes redshifted as it escapes the intense gravity.

By carefully measuring this redshift, scientists can infer several important characteristics of black holes.

1. **Determining the Mass of a Black Hole:** The degree of redshift depends on the **strength of the gravitational field**, which is directly related to the **mass** of the black hole. A stronger gravitational pull produces a greater redshift. Thus, analyzing the redshift of nearby stars or accretion disk emissions helps astronomers estimate the black hole's mass.
2. **Measuring the Spin of a Black Hole:** In systems where gas and dust orbit a black hole, light from the accretion disk becomes gravitationally redshifted as it climbs out of the gravitational well. By examining the **redshifted X-ray spectrum**, particularly the broadened iron emission lines, astronomers can determine the **inner edge** of the accretion disk. Since this inner edge depends on the black hole's **spin**, redshift measurements provide a way to estimate its rotational speed.
3. **Probing the Geometry of the Accretion Disk:** The **pattern of redshift** across an accretion disk offers clues about its **structure and illumination**. Variations in redshift across the disk can reveal how matter moves, how the disk is shaped, and how it is affected by the intense radiation and gravity near the event horizon.
4. **Searching** for **Primordial Black Holes:** Gravitational redshift also plays a role in studying the **evolution of black holes across cosmic time**. By examining how the rate of black hole mergers changes with cosmological redshift, scientists can search for signs of **primordial black holes**—ancient black holes that may have formed shortly after the Big Bang.
5. **Test general relativity** under **extreme conditions** and study the formation and evolution of black holes and the large scale-structure.

Formula for Gravitational Redshift

For light emitted at a distance r from a non-rotating (Schwarzschild) mass M, the gravitational redshift is given by:

$$z = (1 - 2GM/rc^2)^{-1/2} - 1$$

where:

z - the gravitational redshift

r - radial coordinate(the distance from the center of the massive body)

M - the mass of the object

G,c - fundamental physical constants

This equation shows that as **r** decreases (closer to a very compact object), the redshift **z** increases. When **r** approaches the Schwarzschild radius $\mathbf{2GM/c^2}$, the redshift tends toward infinity, meaning light becomes infinitely redshifted.

For rotating black holes, gravitational redshift has no single simple expression. It depends on the black hole's spin, the position of the source, and the direction of emission. In addition to gravity, frame dragging and relativistic motion modify the observed frequency of light.

For a rotating (Kerr) black hole, gravitational redshift becomes more complex due to frame dragging and depends on the observer's position and motion. However, for simplicity, the Schwarzschild expression provides a good approximation at large distances.

2. Gravitational Lensing

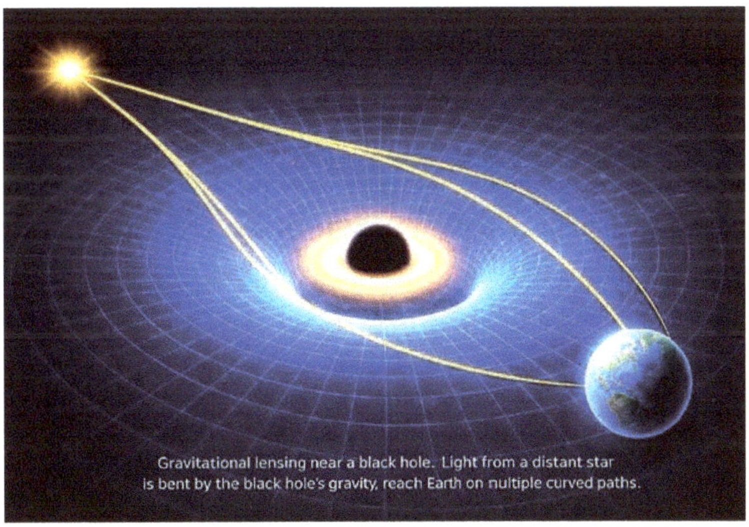

Figure 6.4: AI-generated illustration of light bending around a black hole due to spacetime curvature (gravitational lensing) for conceptual purposes

Gravitational Lensing is the bending of light caused by massive celestial objects such as black holes, neutron stars, and galaxies. This remarkable phenomenon acts like a **natural cosmic telescope**, magnifying and distorting the light from distant sources. Depending on the alignment, it can produce **multiple images**, **bright arcs**, or even a perfect **"Einstein Ring."**

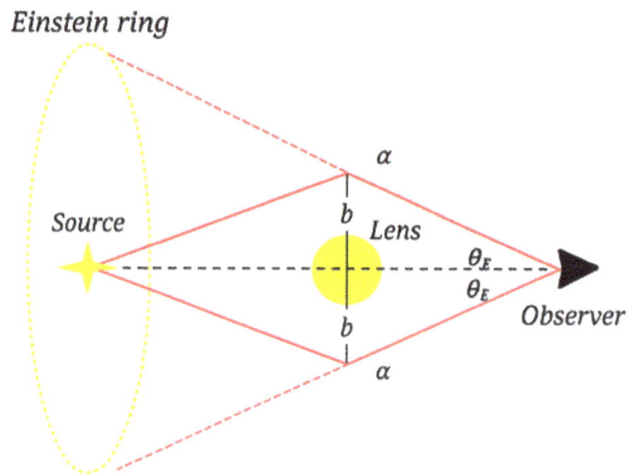

Figure 6.5: AI-generated illustration of Einstein ring for conceptual purposes

An **Einstein ring** is formed when a distant source, a massive intervening object (the lens), and the observer are perfectly aligned in a straight line. The gravitational field of the lens bends the light coming from the background source in all directions around it. Because the bending is symmetrical, the observer receives the deflected light as a complete circle of images, known as an Einstein ring. This circular distortion is a direct consequence of gravitational lensing predicted by Einstein's general theory of relativity which states that mass and energy warp the fabric of spacetime. As a result, light does not always travel in a straight line but follows the curved path created by this warping.

Besides being a fascinating demonstration of relativity, gravitational lensing has become a powerful tool in modern astrophysics. It allows scientists to detect and study objects that would otherwise remain invisible.

How Gravitational Lensing Helps in the Study of Black Holes?

1. **Detecting Hidden Black Holes:** Lensing can reveal the presence of **isolated or dormant black holes** that lack visible accretion disks and are therefore undetectable by conventional methods.

   In a recent study led by researchers from **Durham University, U. K.**, astronomers used gravitational lensing to discover an **ultramassive black hole**— approximately **30 billion times the mass of the Sun**, located in a galaxy hundreds of millions of light-years from Earth.

   This was the **first black hole ever detected using lensing alone**, marking a breakthrough that could help future large-scale telescopes explore even more distant and hidden black holes.

2. **Studying Accretion Disks:** The lensing effect can **distort the light** from an accretion disk around a black hole. By analyzing these distortions, astronomers can learn about the **structure, dynamics, and motion** of material swirling near the event horizon.

3. **Measuring Mass and Spin:** The amount of light bending helps scientists estimate a black hole's **mass** and sometimes even its **spin**. More massive or rapidly rotating black holes cause stronger lensing effects.
4. **Observing Black Hole Jets:** Lensing can **magnify light from relativistic jets** ejected by supermassive black holes, making it possible to study these energetic outflows across immense cosmic distances.
5. **Dark Matter:** Lensing provides strong evidence for **dark matter**. The visible matter in galaxies and clusters is insufficient to account for the observed bending of light, implying the presence of invisible mass.
6. **Dark Energy:** By mapping how light is bent across large scales, lensing helps scientists refine measurements of **dark energy**, the mysterious force driving the accelerated expansion of the universe.
7. **Exoplanets:** An exoplanet is a planet that orbits a star beyond our Solar System. The microlensing technique detects exoplanets by observing brief, subtle increases in a background star's brightness, caused by the gravitational bending of light when a planet and its host star pass in front of it. This method is especially useful for discovering distant and low-mass planets that are otherwise difficult to detect.
8. **Extreme Cosmic Events:** By analyzing lensed light from supernovae, kilonovae, and gamma-ray bursts, astronomers gain valuable insights into neutron star mergers, black hole interactions, and the energetic processes that shape the universe.

Microlensing → Temporary Brightening Without Visible Distortion

Microlensing is a special case of gravitational lensing in which a relatively small object, such as a star or planet, briefly magnifies the light of a background source without producing visible image distortion, unlike gravitational lensing by massive objects that creates multiple or distorted images.

Remarkably, gravitational microlensing has enabled the detection of a possible extragalactic planet associated with **PA-99-N2** in the **Andromeda Galaxy.** In addition, numerous exoplanets, such as OGLE-2003-BLG-235Lb, have been discovered through this method.

9. **Deflection angle for light passing a point- mass object**

$\boldsymbol{\alpha = 4GM/c^2b}$

where

α - deflection angle of light ray

M - mass of lensing object.

B - the impact parameter(closestapproach distance of the light ray)

G, c - gravitational constant and speed of light

This formula shows how much ray of light bends when it passes near a massive object. The greater the mass M or closer the light passes, the stronger the bending of light.

This formula applies in the

Weak gravitational field limit

It assumes a point-mass

(Schwarzschild point lens)

The deflection angle is small

In this regime, rotation effects are negligible, and a Kerr black hole bends light almost exactly like a Schwarzschild (non-rotating) one.

Gravitational lensing is one of the most powerful tools in modern astronomy. When traditional observation methods fail, lensing allows scientists to **detect, magnify, and study** distant and hidden objects—from **black holes** and **exoplanets** to **dark matter** and **cosmic explosions**. It not only confirms the predictions of Einstein's relativity but also continues to open new windows into the mysteries of the cosmos.

3. Time Dilation Near a Black Hole

Time dilation is a fundamental prediction of Einstein's theory of general relativity. It refers to the phenomenon where time passes at different rates for observers in different gravitational fields or states of motion. Near a black hole, time dilation becomes an extreme and measurable effect due to the intense warping of spacetime caused by the black hole's immense gravitational field.

As an object approaches a black hole, the curvature of spacetime increases dramatically. To a distant observer, time for the infalling object appears to slow down. This effect becomes most pronounced near the event horizon, time dilation becomes so extreme that, from the point of view of a distant observer, time seems to stop entirely for the object approaching the black hole. The object appears "frozen" in space, eternally hovering at the edge.

This is not merely an optical illusion, but a genuine physical effect rooted in the nature of spacetime. Inside this warped geometry, time passes more slowly the closer one is to the massive object.

The Degree of Time Dilation Depends On Two Main Factors:

1. **The mass of the black hole:** More massive black holes have stronger gravitational fields and therefore cause more pronounced time dilation.

2. **The distance from the event horizon:** The closer an observer is to the Schwarzschild radius, the slower time passes relative to a distant observer.

Time Dilation Formula

For a non-rotating, spherically symmetric black hole, the time dilation experienced by a stationary observer at a radial coordinate compared to a distant observer is given by

$$t_r/t = (1 - R_s/r)^{1/2}$$

Where

tr - the time interval measured by a clock at radial distance **r**

t - the time interval measured by distant observer (infinity)

r - the radial coordinate measured from the center of a black hole

Rs - Schwarzschild radius(**$R_s = 2GM/c^2$**)

As the radial distance "**r**" approaches the Schwarzschild radius **R_s**, the quantity **$1- R_s/r$** approaches zero, causing the time dilation factor to approach **zero.** This means that time measured by a clock near the event horizon slows dramatically relative to a distant observer.

Example: Sagittarius A*

Consider Sagittarius A*, the supermassive black hole at the center of our galaxy. Under idealized and purely hypothetical conditions, if an astronaut could remain extremely close to its event horizon, time for them would pass far more slowly than for a distant observer. In such an extreme gravitational environment, a **single minute** experienced near the black hole could correspond to **hundreds** or even **thousands of years** passing for observers far away, depending on how close the observer is to the event horizon and on their state of motion. As a result, someone near the black hole would age much more slowly and, when compared with distant observers, would effectively emerge far into the future.

A famous example is the fictional depiction in the movie **Interstellar**, where **one hour** near a black hole corresponds to about **seven years on Earth.**

This dramatic effect arises from gravitational time dilation, a fundamental prediction of general relativity.

Time Dilation - Special Relativity

Time Dilation Formula

$$t = t_0 / (1- v^2/c^2)^{1/2}$$

Where

t- = time interval measured by the stationary observer (time dilated)

t_0 - the time measured by moving clock

v - the relative velocity between observer and the moving object

c - the speed of light

When an object moves at very high speed, its clock runs more slowly compared to a stationary observer. This slowing of time is called time dilation.

Special relativistic time dilation arises from **high velocity**, while gravitational time dilation arises from **strong gravity**; both effects are essential consequences of Einstein's theory of relativity and together govern how time flows in the universe.

In real-world scenarios such as **GPS satellites** both types of time dilation effect and must be accounted for because the satellites'high speed causes time to slow but their weaker gravitational field causes time to speed up.

Engineers use both calculations to ensure their precise timing works correctly.

4. Spaghettification

Black holes exhibit several extraordinary physical effects, among which spaghettification is one of the most dramatic. Also known as the noodle effect, spaghettification describes the extreme stretching an object undergoes when subjected to intense tidal forces near a black hole.

These tidal forces arise because gravity does not act uniformly across an extended object. As something falls toward a black hole, the portion closer to the black hole experiences a much stronger gravitational pull than the portion farther away. This difference in gravitational acceleration causes the object to be stretched along the radial direction while being compressed sideways. The combined effect elongates the object into a thin, filament-like shape, reminiscent of a strand of spaghetti.

Contrary to popular belief, spaghettification does not occur only after crossing the event horizon. Around stellar-mass black holes, the tidal gradient is extremely steep, meaning that significant stretching can begin well outside the event horizon. In such cases, an infalling object would be torn apart before ever reaching that boundary.

In contrast, supermassive black holes, with masses millions to billions of times greater than the Sun, possess vastly larger event horizons and much weaker tidal gradients at those distances. As a result, an object—or even a human in principle—could cross the event horizon of a supermassive black hole without immediately experiencing noticeable spaghettification. The destructive tidal forces would only become dominant much deeper inside, closer to the central singularity.

Figure 6.6: Spaghettification

Image Credit: NASA/Imagine the Universe!

In essence, spaghettification is a striking and graphic illustration of the extreme nature of gravity near black holes—a phenomenon where matter, no matter how rigid or durable, is ultimately torn apart and stretched by the overwhelming warping of spacetime curvature.

Concluding Remarks

The physical characteristics of black holes reveal a realm where the laws of physics are pushed to their absolute limits. Quantities such as extreme density, rapid spin, and intense gravitational fields define these objects not merely as collapsed stars, but as profound distortions of space and time itself. Unlike ordinary celestial bodies, black holes possess properties—such as entropy and temperature—that connect gravity with thermodynamics, hinting at a deeper, unified description of nature.

The observable effects surrounding black holes—gravitational redshift, lensing, and time dilation—demonstrate how strongly they influence their environment, altering the very fabric through which light and time propagate. More dramatic phenomena, such as spaghettification, illustrate the extreme consequences of tidal forces near the event horizon, where classical intuition breaks down entirely.

Taken together, these characteristics show that black holes are not merely astronomical objects, but natural laboratories for testing the principles of General Relativity and exploring the boundary between classical physics and quantum theory. They challenge our understanding of space, time, and information, and continue to play a central role in modern astrophysics.

Thus, the study of black holes is not only about understanding the most extreme objects in the universe, but also about uncovering the fundamental laws that govern reality itself.

Chapter 7
Extraordinary Black Holes and Their Systems

Introduction

Today, black holes are not only well-established astrophysical entities but also common components of the universe, found in stellar systems, galactic centers, and even in merging cosmic structures.

Among the countless black holes known or suspected, certain ones stand out as truly extraordinary because of their immense masses, great distances, powerful jets, intense accretion of matter, and exceptionally luminous emissions. Some lie in the farthest reaches of the observable universe, while others grow by consuming surrounding matter at tremendous rates, shining with extraordinary brightness. These remarkable systems provide valuable insights into the diverse nature of black holes and deepen our understanding of their formation, evolution, and influence on the universe.

Among the countless black holes known or suspected, certain ones stand out as special due to their extraordinary features. These include:

The Most Massive Confirmed Black Hole: TON 618

TON 618 is one of the most extraordinary quasars ever discovered and is believed to host the most massive confirmed black hole known to science. It represents the extreme limits of black hole growth, luminosity, and energy production in the universe.

Origin of the Name: TON 618

The designation TON 618 comes from the Tonantzintla Observatory in Mexico, from which the prefix "TON" is derived. The number 618 indicates that it was the 618th object listed in a catalogue compiled there.

Discovery of TON 618

In **1957** Mexican astronomers **Braulio Iriarte** and **Enrique Chavira** first cataloged TON 618 as an unusually bright object during a sky survey at the **Tonantzintla Observatory**.

At that time, it was not known to be a quasar or related to a black hole and was simply listed as a peculiar, very bright source.

In **1970,** an Italian astronomer detected **strong radio emissions** from TON 618. This discovery showed that it was not an ordinary star but a highly energetic and distant cosmic object, pointing toward its true nature as a quasar.

Later Studies: With improved observations, astronomers realized that the extraordinary luminosity of TON 618 could only be explained if it were powered by a supermassive black hole actively accreting matter.

Mass of the Black Hole

Astronomers estimated the mass of the black hole in TON 618 by studying the motion of gas in its accretion disk and the broad emission lines in its spectrum. These measurements revealed that the central black hole has a mass of about **66 billion M⊙,** making it one of the most massive black hole ever discovered.

Distance: TON 618

TON 618 is extremely distant. The light we see today left the quasar when the universe was much younger, about **10–11 billion years** ago. However, because the universe has been expanding while the light traveled toward us, the present-day distance is much larger. When cosmic expansion is taken into account, its **comoving** distance exceeds **18 billion light - years.**

Location

TON 618 lies at the center of a very distant host galaxy whose individual structure and name are not known because the quasar's extreme brightness dominates the light of the galaxy. TON 618 appears in the direction of the constellations **Canes Venatici** and **Coma Berenices,** whose stars belong to the Milky Way, but the quasar itself is located in a very distant host galaxy.

Immense Size and Power

The physical scale and energy output of TON 618 are almost unimaginable:

Event horizon radius: $\approx 1.9 \times 10^{14}$ **meters**

Matter consumption rate: $\approx 10^{23}$–10^{24} **Kg/s.**

Energy output (luminosity): $\approx 4 \times 10^{40}$ **watts**

Brightness comparison: roughly **100–140 trillion times the luminosity of the Sun**

Surrounding Emission Region: Nebula

A luminous nebular structure extending up to ~ **330,000 light-years,** nearly three times the diameter of the Milky Way. The brilliant accretion disk around the black hole, composed of superheated gas and dust spiraling inward, produces intense radiation. This makes TON 618 one of the brightest objects in the observable universe despite its enormous distance.

The TON 618 stands as a powerful example of how supermassive black holes can grow to truly cosmic scales, even in the early universe. Its discovery has provided crucial insight into the formation and evolution of the largest black holes.

After TON 618, one of the most massive well-confirmed supermassive black holes known is in **Holmberg 15A,** with current mass estimates ranging from about **22 to 40 billion times that of the Sun.** It is located about **700 million light-years** from Earth, in the galaxy cluster **Abell 85,** in the constellation **Cetus.**

Most Massive Black Hole If Confirmed: Phoenix A

Phoenix A is a supermassive black hole whose mass is estimated to be approximately **100 billion M$\odot$.** If this value is confirmed by future observations, it would make Phoenix A the most massive black hole ever discovered. At present, its mass is based on theoretical models and indirect measurements, so it is still considered uncertain. .

Location

Phoenix A is a true cosmic giant located at the heart of the Phoenix Galaxy Cluster. The light we observe today left it about **5.7 billion years** ago, while its present-day (comoving) distance—accounting for the expansion of the universe—is approximately **8.5 - 9 billion light-years** and is thought to have grown over billions of years through galaxy mergers and by consuming enormous amounts of surrounding matter.

Fascinating Features of Phoenix A

1. Phoenix A hosts an extremely massive supermassive black hole making it about **24,000** times more massive than **Sagittarius A*** at the center of the Milky Way.
2. The Schwarzschild radius of the Phoenix black hole is approximately **295 billion kilometers (≈ 2,000 AU).**
3. The black hole is accreting matter at an estimated rate of roughly **3×10^{24} - 4×10^{24} kilograms per second,** roughly **60 solar mass per year,** powering one of the most luminous active galactic nuclei known within a galaxy cluster environment.

The black hole is believed to be **rotating,** although the exact speed of its spin is still unknown.

Its immense gravity likely powers powerful jets of radiation and particles, which are responsible for energetic activity seen in **active galactic nuclei (AGN).**

The jets are not just destructive. They stir and heat the surrounding gas, triggering a cycle that affects star formation. Because of this ongoing process, the Phoenix Cluster serves as an unique laboratory for studying galactic evolution and how galaxies change over time.

Discovery: The Phoenix Cluster (SPT-CLJ2344-4243) was identified in survey data from the **South Pole Telescope** and brought to prominence in 2012 through a study led by **Michael McDonald** and collaborators. The results, published in *Nature,* revealed it to be one of the most massive and X-ray luminous galaxy clusters known.

The discovery was remarkable because the Phoenix Cluster turned out to be one of the hottest and most massive known galaxy clusters in the universe.

Phoenix A: A Laboratory of Extreme Physics

Phoenix A is not just a black hole; it is a laboratory for studying the extreme limits of gravity, mass, and energy. If its estimated mass is confirmed, it will redefine what astronomers consider possible for black hole growth and will stand as the most massive black hole ever known.

Meaning of the Name "Phoenix": The name Phoenix comes from ancient mythology, where the phoenix is a legendary bird that burns itself and is reborn from its own ashes, symbolizing renewal, power, and immortality.

This name beautifully reflects the nature of Phoenix A and its host cluster—constantly transforming matter and energy, reshaping its cosmic environment.

Figure 7.1: AI-generated Illustration of comparison of Phoenix A if confirmed and TON 618 for conceptual purposes

Lowest-Mass Stellar-Mass Black Hole : GRO J0422+32

GRO J0422+32 is a fascinating binary system composed of a confirmed black hole and a low-mass companion star. It is classified as a low-mass X-ray binary and an X-ray nova, as it occasionally erupts in powerful X-ray outbursts when matter from the companion spirals inward toward the black hole's invisible core.

Companions of System

The black hole in the system is considered as one of the **lowest stellar-mass black hole** candidates known, with an estimated mass of approximately **3.7–5.0 M⊙**.

Its companion is an **M-type red dwarf** with a mass of about **0.3–0.5 M⊙** and a radius roughly **30–50% that of the Sun.**

The binary system has an exceptionally short orbital period of only about **5.1 hours,** permitting detailed analysis of its orbital dynamics and leading to more accurate estimates of the component masses. Because its estimated mass lies near the proposed "mass gap" between neutron stars and black holes, GRO J0422+32 remains an important object in studies of compact object formation.

Location: The system is located in the **constellation Perseus** within the Milky Way, at a distance approximately **7,800–8,100 light-years** away from Earth.

Discovery: The system was first detected during an X-ray outburst in 1992 by NASA's Compton Gamma Ray Observatory. Subsequent optical observations established it as a low-mass black hole candidate, making it one of the smallest stellar-mass black holes known. The discovery was published in *Astro Physical Journal (413, L21), in 1993*.

Lowest Mass Stellar-mass Black Hole Candidates

The identity of the lowest stellar-mass black hole remains one of the most debated questions in black hole astronomy. Over the years, different observations and studies have proposed several candidates for this distinction. Objects such as GRO J0422+32, once regarded as leading examples, are now joined by newer contenders, keeping the subject open to discussion. As a result, no single object can yet be identified with complete certainty as the lowest confirmed stellar-mass black hole.

Competitors

1. **XTE J1650-500:** One of the leading low-mass black hole candidates, XTE J1650-500 has been estimated to have a mass of about **4–7 times that of the Sun,** though some studies suggest it could be closer to the lower end of this range. Because its exact mass remains uncertain, it cannot yet be confirmed as the lowest-mass stellar black hole.

XTE J1650-500 is located in the constellation Ara, within our Milky Way galaxy, at an estimated distance of **about 8,000–10,000 light-years** from Earth. Its event horizon would span only a few tens of kilometers, highlighting how such compact objects can contain several times the Sun's mass within an extremely small region.

2. **2 MASS J05215658+4359220:** It is located in the constellation **Auriga,** about **3,000 light-years** from Earth, within our Milky Way galaxy. The star appears to form a binary system with an unseen low-mass black hole candidate.

The system drew major attention in 2019 when astronomers found evidence that the visible giant star may be orbiting an unseen companion with a mass of about **3.3 times that of the Sun**. If confirmed as a black hole, it would be one of the lightest stellar-mass black holes ever discovered, lying in the poorly understood mass gap between neutron stars and black holes. This quiet, non-accreting system is especially important because it suggests that many hidden black holes may exist in our galaxy without emitting bright X-rays.

[Meaning of **2MASS J05215658+4359220:**

2 MASS = **Two Micron All-Sky Survey,** J = coordinates are given in the J2000.0 celestial reference system, 05215658 = right ascension (05h 21m 56. 58s), +4359220 = declination (+43° 59′ 22. 0″)]

Fastest-Spinning Stellar-Mass Black Hole: GRS 1915+105

GRS 1915+105 is the name of the binary system containing a black hole and a low-mass companion star. The black hole in the system holds the record as the **fastest-spinning stellar-mass black hole** ever measured, astonishingly close to the theoretical maximum. This corresponds to an effective horizon rotation frequency of a few hundred times per second for a stellar-mass black hole.

Location: The GRS 1915+105 system is located in the Milky Way, in the constellation.

Aquila, at a distance of approximately **36,000 light-years** away from Earth.

Black hole: The black hole in the system has an estimated mass of approximately **9–18 M☉**. Its dimensionless spin parameter is extremely high, with estimates of **$a^* \gtrsim 0.98–0.999$,** indicating rotation very close to the theoretical maximum allowed by general relativity. It rotates at over 1,000 times per second, approaching the theoretical maximum speed. The black hole rotates so rapidly that the very fabric of spacetime near the event horizon is forced to swirl at nearly the speed of light.

Companion: The companion star is a low-mass K/M-type giant, with an estimated mass of approximately **0.75–1.3 M☉,** orbiting the black hole with a period of about **33.5 days.**

Discovery: The system was first detected in 1992 by NASA's Compton Gamma Ray Observatory as a gamma-ray transient. Data from NASA's Chandra X-ray Observatory and the Rossi X-ray Timing Explorer (RXTE) were used to model this inner region, revealing its near-maximal rotation. The discovery paper was published in *Nature Astronomy (2023).*

Special Features

1. **X-ray "Heartbeat" Oscillations:** The black hole in GRS 1915+105 exhibits remarkable quasi-periodic X-ray variability, including rhythmic oscillations with timescales of roughly **tens of seconds (≈50 seconds in certain states).** This repeating pattern has been described as a "cosmic heartbeat," making the system one of the most intensively studied black holes in the Galaxy.

2. **Relativistic Jets (Microquasar Nature):** The system launches powerful **relativistic jets**—narrow streams of charged particles traveling at velocities close to the speed of light. Because it produces scaled-down quasar-like jets within our own galaxy, GRS 1915+105 is classified as a **microquasar.**

3. **Extreme X-ray Variability:** It displays dramatic and rapid X-ray variability across multiple timescales, reflecting instabilities in the inner accretion disk near the event horizon.

4. **Long Orbital Period:** With an orbital period of approximately **33.5 days**, it possesses one of the longest known periods among **low-mass X-ray binaries (LMXBs).**

Together, these extraordinary properties make GRS 1915+105 one of the most dynamic and remarkable stellar-mass black hole systems in the Milky Way galaxy.

Fastest-Spinning Supermassive Black Hole: NGC 1365

NGC 1365 is a giant barred spiral galaxy, located about **56–61 million light-years** away in the constellation **Fornax.** It is also known as the Fornax Propeller Galaxy or the Great Barred Galaxy.

Black Hole: At its core lies a supermassive black hole with a mass of approximately **2 million M⊙.**

Spin of the Black Hole: Nearly the maximum possible value, with a dimensionless spin parameter **a* ≈ 0.97**, indicating rotation extremely close to the relativistic limit. It is generally recognised as one of the fastest - spinning at approximately 84% of the speed of light. This implies that spacetime at the event horizon is dragged around at roughly half the speed of light. Such "fastest spin" claims are model-dependent and inferred from X-ray spectroscopy rather than directly measured. Even a seemingly slow rotation period therefore corresponds to enormous linear velocities at the event horizon.

Nature: The black hole powers an **active galactic nucleus (AGN)**, releasing enormous energy as it accretes gas and dust from its surroundings. The high spin rate suggests that this black hole grew through a process of **ordered accretion** — a steady inflow of gas and dust — rather than through chaotic galaxy or black hole mergers. **Discovery**: Its spin was measured using data from ESA's XMM-Newton and NASA's NuSTAR observatories, with results published by Guido Risaliti and colleagues in *Nature* (2013) and recognised as a "Fastest Spinning Supper Massive Black Hole."

Lowest Spinning Supermassive Black Hole

Although **Sagittarius A*** was once considered a candidate for the lowest-spinning supermassive black hole; recent studies indicate that its event horizon rotates at about **60%** of the maximum possible rate. Therefore, no truly low-spin supermassive black hole has yet been reliably identified.

Lowest-Spinning Stellar-Mass Black Hole: A0620-00

The system A0620-00 is widely regarded as containing one of the lowest-spinning stellar-mass black holes identified to date. The system lies in the constellation **Monoceros** in the Milky Way galaxy**,** close to the celestial equator, which allows it to be observed from both hemispheres. It is located about **3,300 - 3,500 light-years** away from Earth.

Discovery: The system was first detected in 1975 by the **Ariel 5 satellite** during a dramatic X-ray outburst. It became one of the brightest X-ray sources in the sky at that time and was initially classified as an **X-ray nova**. The discovery was reported by **Martin Elvis** and collaborators and published in *Nature* (1975).

Binary System Components

The black hole in A0620-00 has a mass of about **6–7 M⊙,** making it one of the best-constrained stellar-mass black holes known. Its event horizon diameter is roughly **38–40 kilometers**.

The companion is a K-type main-sequence star with a mass of about **0.4–0.6 M⊙** and a radius approximately **0.6–0.7 times that of the Sun**. Its surface temperature is around **4,400–4,900 K**, giving it a characteristic orange appearance.

The two objects orbit each other every **7.75 hours,** an exceptionally **short orbital period** that allows precise measurement of the system's dynamical properties.

The intense gravity of the black hole distorts the companion star into an **ellipsoidal shape,** causing regular brightness variations as seen from Earth.

A0620–00 has shown at least two major X-ray outbursts, the most famous in **1975,** when it briefly became one of the brightest X-ray sources ever observed. During quiet phases, it remains faint but still shows signs of weak accretion and **occasional jet activity.**

Spin of the Black Hole: The black hole in A0620-00 is currently regarded as **one of the lowest-spin stellar-mass black holes** measured, with a dimensionless spin parameter of approximately **$a^* \approx 0.12$** (with uncertainties allowing values close to zero).

The spin was determined using the continuum-fitting method, which models the thermal X-ray emission from the inner accretion disk. Observations from missions such as the Rossi X-ray Timing Explorer and the Chandra X-ray Observatory, combined with modern relativistic accretion-disk models, enabled this measurement.

Scientific Importance: A0620–00 is vital for testing black hole physics because it shows that high spin is not essential for accretion or energy release. It challenges models that link powerful jets exclusively to rapid rotation and suggests that some black holes may be born with low spin, inherited from slowly rotating parent stars.

Accretion and Jet

Despite its slow spin, A0620–00 still shows accretion and weak jets, proving that spin alone does not control jet formation. It therefore serves as a natural laboratory for studying the interplay between spin, accretion, magnetic fields, and jets.

Most Distant and Earliest Known Black Hole Candidate: UHZ1

The **Ultra–High–Redshift source1**(UHZ1) is a very distant, early-universe galaxy, located at a redshift of about **$z \approx 10$** (very early cosmic time). It is known for hosting a rapidly growing supermassive black hole.

Cosmic Epoch: With a redshift of **$z \approx 10$,** UHZ1 is observed as it existed when the universe was only about **470 million years** old. It is therefore among the most distant and earliest known accreting black hole candidates identified to date.

The light we detect today has traveled for approximately **13.2 billion years,** meaning we see UHZ1 as it appeared shortly after the cosmic dawn.

Black Hole: Based on its strong X-ray emission and accretion modeling, the central black hole is estimated to have a mass of roughly **10–100 million M⊙.** Such a large mass at such an early epoch suggests that black holes in the young universe either formed from massive initial "seed" objects or grew extremely rapidly through efficient accretion.

Detection: UHZ1 is observed behind the massive galaxy cluster **Abell 2744**. Its detection was made possible through **gravitational lensing**, in which the immense gravity of the foreground cluster magnifies the light from this distant background source.

Discovery: The object was identified using data from NASA's Chandra X-ray Observatory, combined with infrared observations from the James Webb Space Telescope. The intense X-ray emission provided strong evidence for the presence of a

rapid accretion in the early universe. The discovery paper was published in *Nature Astronomy (2023).*

Scientific Significance: The discovery of UHZ1 demonstrates that massive black holes already existed less than half a billion years after the Big Bang. This finding challenges conventional models of black hole formation and growth, suggesting either the presence of unusually massive early seed black holes or sustained periods of rapid accretion in the early universe.

UHZ1 is currently regarded as one of the earliest and most distant known black hole candidates, although future observations may reveal even earlier examples.

Second Most Distant Known

Black Hole: CAPERS-LRD-z9

In addition to UHZ1 ($z \approx 10$), other extremely high-redshift systems such as **CAPERS-LRD-z9** have also been identified as potential hosts of rapidly growing black holes in the early universe. The mass of the supermassive black hole in the system is estimated to be approximately **300 million M⊙.**

CAPERS-LRD-z9 has a redshift of approximately **$z \approx 9$,** meaning its light has traveled for about **13.2–13.3 billion years**. It is observed when the universe was roughly **500 million years** old, about **30 million years** later in cosmic time than UHZ1.

Although slightly younger in cosmic epoch than UHZ1, CAPERS-LRD-z9 likewise demonstrates that substantial black hole growth was already underway within the first half-billion years after the Big Bang.

Together, these discoveries indicate that supermassive black holes began forming and accreting matter remarkably early in the history of the universe.

Nearest Known Black Hole and Nearest Known Black Hole Binary System

Gaia BH1

Gaia BH1 is one of the most remarkable recent discoveries. It is a binary system composed of a stellar-mass black hole and a Sun-like star, located approximately **1,560 light-years** away in the constellation Ophiuchus. It is currently regarded as the closest known quiescent stellar-mass black hole system to our Solar System.

The black hole itself is invisible, as it does not actively accrete matter. Its presence was inferred through precise measurements of the orbital motion of its companion star, which moves around an unseen massive object.

Binary System Details

The black hole has an estimated mass of approximately **9.6-10 M⊙**.

Its companion is a **G-type main-sequence star,** similar to the Sun, with a mass of about **0.93 M⊙** and a radius of roughly **0.99 times that of the Sun.**

The two objects orbit their common center of mass approximately every **185.6 days**.

Because Gaia BH1 is not currently accreting material and emits no detectable X-ray radiation, its **spin parameter remains unknown.**

This wide orbit and the Sun-like nature of the companion, make Gaia BH1 very different from typical X-ray binary systems, which usually contain hot, massive stars and tight orbits.

Discovery and Confirmation

Gaia BH1 was discovered in **2022** by a team led by **Dr. Kareem El-Badry,** using high-precision data from ESA's Gaia Space Observatory. The presence of the black hole was later confirmed by spectroscopic observations from the Gemini North Telescope on Mauna Kea, Hawaii. The discovery paper was published in Monthly Notices of the *Royal Astronomical Society (2022).*

Despite being the nearest known black hole, Gaia BH1 poses absolutely **no danger to Earth**. At about **1,560 light-years** away, it is far too distant to affect our Solar System in any physical way. It is important purely for its scientific value.

The second-closest known black hole is **Gaia BH3,** located at about **1,926 light- years** from Earth.

Name and Symbolism

In Greek mythology, *Gaia* is the primordial goddess of Earth, symbolizing life and creation. In science, the Gaia mission represents a grand effort to map and understand the structure, motion, and relationships of objects in our galaxy.

A Young Black Hole: W49B

W49B is a supernova remnant located in the constellation Aquila, approximately **25,000–26,000 light-years** from Earth within the Milky Way. The explosion that created black hole is estimated to have occurred roughly **1,000 years** ago, making it one candidate of the youngest known supernova remnants in our galaxy so far.

Unlike many supernova remnants, W49B shows no clear evidence of a surviving neutron star at its center. This absence has led some researchers to propose that the explosion may have resulted in the direct formation of a black hole. If confirmed, it would represent one of the youngest known black hole candidates.

Discovery: In 2013, a study using NASA's Chandra X-ray Observatory suggested that W49B might have formed a black hole rather than a neutron star.

Conclusion: W49B is recognized as a unique, likely black-hole-producing, young supernova remnant, but it remains a subject of intense, active research, rather than a definitively confirmed black hole.

The Brightest and Fastest-Growing Quasar: QSO J0529–4351

J0529–4351 is among the most luminous quasars ever observed and hosts one of the most extreme supermassive black holes identified to date. Its extraordinary brightness is powered by an immense accretion rate, making it one of the most energetic known objects in the universe.

Location: It lies in the constellation **Pictor** at a distance of approximately **12 billion light-years** away from Earth.

Black Hole Properties

The central black hole has an estimated mass of roughly **15–20 billion solar masses** ($\approx 1.5–2.0 \times 10^{10}$ M⊙). Although its spin has not yet been directly measured, its extreme luminosity indicates a highly efficient accretion process.

The quasar is surrounded by an enormous accretion disk extending several light-years across, placing it among the **largest known accretion structures around any black hole.**

Luminosity: Its total luminosity is estimated to be approximately **5×10^{14} times that of the Sun**, making it one of the brightest persistent sources in the observable universe

Current models indicate that J0529–4351 accretes roughly hundreds of solar masses of gas per year—approximately **one solar mass per day.** Not all of this material is added to the black hole's mass; a significant fraction is converted into energy and emitted as radiation through highly efficient accretion processes. As a result, the net growth rate of the black hole is somewhat lower than the total accretion rate, though it still ranks among the most rapidly growing supermassive black holes known.

This extraordinary rate of mass inflow and energy conversion is what makes J0529–4351 one of the most luminous quasars ever observed.

QSO J0529–4351 holds the distinction of being the **brightest quasar** currently known and one of the most rapidly growing supermassive black holes ever identified. No other observed quasar matches its extraordinary luminosity or the remarkable rate at which it accretes matter.

Why is it So Bright

QSO J0529–4351 shines with such intensity because of its enormous, superheated accretion disk. Gas falling toward the black hole becomes compressed and heated to millions of degrees, releasing radiation across the electromagnetic spectrum—especially ultraviolet and X-rays.

This process makes the quasar brighter than entire galaxies, making it one of the most visible beacons in the early universe.

Discovery and Identification

J0529–4351 was first recorded in 1980 during surveys conducted with facilities operated by the European Southern Observatory (ESO).

In 2023, detailed spectroscopic analysis and modern observational data confirmed its identity as one of the most luminous quasars known, revealing the extraordinary mass of its central black hole and its extreme accretion rate.

A Cosmic Puzzle: J0529–4351 is extraordinarily luminous and highly active, even when compared with other extremely massive black holes. Its exceptional brightness may result from a combination of dense surrounding gas, highly efficient accretion, and possibly favorable magnetic-field and accretion-disk structures.

As one of the most luminous quasars ever observed, J0529–4351 serves as a key object for understanding quasar physics, extreme accretion processes, and the growth of supermassive black holes in the universe.

One of the Most Powerful Known Jet System: Porphyrion

The Porphyrion jet system represents one of the most powerful and largest twin relativistic jet structures ever observed. Its total energy output rivals that of trillions of Suns, and its jets extend across an astonishing **23 million light-years**—a length equivalent to lining up roughly **140 Milky Way galaxies** end to end.

In astrophysics, a pair of jets refers to two narrow, oppositely directed streams of high-energy plasma launched from the immediate vicinity of a supermassive black hole. These jets travel at velocities close to the speed of light and can extend far beyond the boundaries of their host galaxies.

Location: Porphyrion's jets originate from the **galaxy J152932. 16+601534. 4,** located approximately **7.5 billion light-years** from Earth. This galaxy is believed to be significantly more massive than the Milky Way, possibly by a **factor of ten.**

Black Hole: At its core lies a supermassive black hole likely containing several **billion solar masses (M⊙),** although its exact mass and spin remain under investigation.

Extraordinary Stability: One of the most remarkable features of the Porphyrion system is the exceptional straightness and stability of its jets. Their near-perfect alignment over a span of **23 million light-years** suggests that the black hole's rotation axis has remained stable for **tens of millions of years**. Such long-term directional stability is rare and may be essential for producing a jet system of this unprecedented scale.

Power of the Jets: Porphyrion's jets have a combined power of approximately **10^{39}- 10^{40} watts**, equivalent to the energy output of **trillions of Suns**. This immense power allows the jets to transport energy across intergalactic distances far beyond the host galaxy.

Discovery: The Porphyrion jet system was identified in **2024 using the Low-Frequency Array (LOFAR)** radio telescope network in Europe. Its existence was later confirmed through multi-wavelength observations with major facilities, including:

W. M. Keck Observatory, England (optical observations)

Giant Metrewave Radio Telescope, India (radio observations)

The findings were published in *Nature on 18 September 2024*, marking a major milestone in the study of large-scale extragalactic jets.

Significance in Modern Astrophysics

The discovery of Porphyrion demonstrates that jets powered by supermassive black holes can remain coherent, stable, and extraordinarily powerful over distances far greater than previously confirmed. It suggests that:

1. Magnetic fields near supermassive black holes may be stronger or more highly organized than previously assumed.
2. Relativistic jets can influence intergalactic environments on truly cosmic scales.
3. Existing models of jet propagation, collimation, and energy transport may require refinement.

Mythological Origin of the Name "Porphyrion"

The name "Porphyrion" comes from the King of the Giants in Greek mythology—an apt title for this colossal, universe-spanning structure.

Another Powerful Contender: J1610+1811

Another extraordinary example of extreme jet activity is the distant quasar J1610+1811, located approximately **11.6 billion light-years** from Earth. Its relativistic jets carry an enormous amount of energy—comparable to a significant fraction of the radiative power emitted by the hot accretion disk surrounding its central supermassive black hole.

The combined kinetic power of its jets is estimated to be on the order of **10^{39}-10^{40} watts.** These jets are highly relativistic, traveling at speeds of roughly **92–98% of the speed of light**. The energy transported by them is equivalent to the total luminosity of **tens of trillions of Suns**, underscoring the extreme nature of this system.

Comparison with Porphyrion

Although the jet powers of Porphyrion and J1610+1811 fall within a similar range, astronomers estimate that Porphyrion's jets may be slightly more powerful overall.

However, Porphyrion is considered exceptional not primarily because of marginally higher instantaneous jet power, but because of:

1. **Gigantic size:** Its jets extend approximately 23 million light-years
2. Immense total energy injection into intergalactic space over time
3. Long duration of sustained activity, allowing continuous energy transport
4. Large-scale environmental impact, influencing intergalactic and possibly cluster-scale structures.

Thus, Porphyrion is regarded as one of the largest and most energetically influential jet systems ever discovered, though not necessarily the one with the highest instantaneous jet power.

A Close Pair of Supermassive Black Holes: NGC 7727

NGC 7727 is a remarkable galaxy that hosts one of the closest known pairs of supermassive black holes discovered to date. It is located approximately **89 million light-years** away from Earth in the constellation **Aquarius**.

Black Holes: The system contains two supermassive black holes separated by only about **1,600 light-years**—an unusually small distance for such massive objects within a single galaxy. The more massive black hole has an estimated mass of about **154 million M⊙,** while the smaller companion has a mass of roughly **6.3 million M⊙**.

Although the two black holes are gravitationally bound, they have not yet formed a tight binary system. Over time, dynamical interactions with surrounding stars and gas are expected to cause them to spiral inward. Astronomers estimate that they may merge in approximately **250 million years**, potentially producing strong gravitational waves.

Discovery: This discovery was led by astronomer **Karina Voggel** of the University of Strasbourg, using data from the European Southern Observatory and observations from the Very Large Telescope in Chile. The findings were published in *Astronomy & Astrophysics in November 2021.*

Scientific Significance: The NGC 7727 system provides valuable insight into how galaxies grow through mergers and how supermassive black holes evolve following galactic collisions. It offers a rare observational snapshot of a late-stage galaxy merger in which two central black holes are approaching eventual coalescence.

Important Black-Hole Advances (January–April, 2026)

1. Candidate For the Closest Pair of Supermassive Black Holes

Markarian 501

Markarian 501 is a galaxy that may host a remarkably close pair of supermassive black holes — a striking recent discovery in the month of April, 2026. Astronomers have found strong evidence that its nucleus contains a very tight binary system that may merge within about **100 years**, an extraordinarily short timescale in cosmic terms. The galaxy is located about **460 million light-years** away in the constellation Hercules.

Observations and Black Holes

Long-term radio observations spanning nearly **23 years** revealed not only the main relativistic jet in Markarian 501, but also evidence of a second jet, providing some of the strongest indications yet that its nucleus may contain two orbiting supermassive black holes. The candidate pair appears to complete an orbit roughly every **121 days**, with an estimated separation of only about **0.004 — 0.009 light-years — astonishingly close for such massive objects.**

Each of the proposed black holes is thought to have a mass ranging from approximately **100 million to 1 billion times that of the Sun,** making them among the most massive close black hole pairs ever suggested. Depending on their true masses and orbital properties, the two black holes could spiral inward and merge in as little as a century.

If future observations from multiple independent sources confirm the discovery in Markarian 501, it may surpass NGC 7727 as the closest known pair of supermassive black holes.

A possible future source of low-frequency gravitational waves observatories like the LISA may detect gravitational waves from merging black holes.

Discovery: The recent discovery of a possible very close pair of supermassive black holes in its core was announced in **April 2026** by an international team led by **Silke Britzen** of the Max Planck Institute for Radio Astronomy, after 23 years of observations. The discovery of a possible ultra-close pair of supermassive black holes in Markarian 501 has been accepted for publication in *Monthly Notices of the Royal Astronomical Society (MNRAS)* and is expected to appear in an upcoming issue.

2. A Black Hole Born Without a Supernova

In February 2026, astronomers reported the likely **direct collapse of a massive luminous star into a black hole in the Andromeda Galaxy.** The star, designated **M31-2014-DS1,** was located about **2.5 million light-years** from Earth and had an estimated mass of roughly **13 M⊙.** Such an event is known as a **"failed supernova,"** or **direct collapse**, in which a star collapses into a black hole without producing the bright explosion normally associated with a supernova.

The quoted mass of about **13 M⊙** likely represents the star's mass during a **late stage of stellar evolution**, possibly the **supergiant phase**, rather than its original mass at birth. Massive stars can lose a significant portion of their mass through **stellar winds and other processes** during their lifetime.

Formation of the Black Hole

The sudden disappearance of the star, together with the detection of infrared emission from surrounding material, strongly suggests that the star underwent direct gravitational collapse into a black hole. Astronomers observed that the luminous star gradually faded from view and eventually vanished in visible light, while faint infrared radiation was still detected from the region. This behavior indicates that the stellar core collapsed and the remaining material was either accreted onto the newly formed black hole or dispersed into the surrounding space. Observations suggest that the star collapsed into a **black hole with a mass of about 5 M⊙** and no visible supernova explosion was detected. If confirmed through continued observations, this event may represent **one of the youngest known black holes** and would provide rare observational evidence of the **birth of a black hole**.

Why No Supernova Occurred

When a massive star collapses directly into a black hole, only part of the stellar material becomes part of the black hole itself. Some of the remaining matter may temporarily form an **accretion disk or surrounding envelope of gas** around the newly formed black hole. At the same time, an enormous amount of energy is carried away by **neutrinos**, particles produced in huge numbers during core collapse. Because the collapsing core fails to generate a sufficiently strong **shock wave**, the outer layers of the star are not violently expelled as in a normal supernova. Instead, the star collapses quietly. Some material may still drift away from the outer layers, while the rest falls inward.

As a result, a progenitor star of about **13 M⊙** leave behind a **black hole** of only about **five M⊙.** Events of this kind are believed to be **rare**, which is why their observation is of great scientific interest.

Discovery: The event was identified using data from **NASA's NEOWISE Mission**, with analysis led by **Dr. Kishalay De of Columbia University.** Astronomers examined observations collected between **2005 and 2023**, along with data from several ground-based and space telescopes. The results were reported in the journal ***Science*** *in February 2026.*

3. Another Failed Supernova: NGC 6946-BH1

Astronomers are also studying another possible example of a failed supernova in the object **NGC 6946-BH1,** located in the spiral galaxy **NGC 6946**, about **22 million light-years** away from Earth. Although strong evidence exists, the final confirmation of that event is still under investigation.

4. Discovery of a Rapidly Growing Supermassive Black Hole Launching Powerful Jets

In January 2026, astronomers reported the discovery of an unusual quasar in the early universe containing one of the **fastest-growing supermassive black holes known relative to its size.** What makes this object particularly remarkable is that the black hole is **accreting matter at an extremely high rate while simultaneously producing powerful jets and intense X-ray radiation.**

What makes this quasar unusual is the simultaneous presence of extremely rapid growth and powerful relativistic jets. In most theoretical models, supermassive black holes accreting at very high rates produce intense radiation but relatively weak jets. Conversely, systems with strong jets usually accrete matter more slowly. The newly discovered quasar appears to combine both behaviors, suggesting that black hole growth in the early universe may occur in ways not fully explained by current theories. Such a combination is extremely rare.

Properties of the Quasar

Location: This exceptional quasar has a **cosmological redshift of** approximately **z ≈ 3.4,** placing it roughly **11–12 billion light-years** away from Earth. At this distance we observe the black hole as it existed when the Universe was only about **two billion years old.**

Black Hole: The mass of the central black hole is estimated to be about **440 million M⊙.** Observations in X-ray wavelengths indicate that the black hole is **accreting matter at roughly thirteen times the Eddington limit.** The Eddington limit normally represents the maximum steady rate at which matter can fall into a black hole before radiation pressure pushes the material away. Accretion at thirteen times this limit suggests that the black hole is consuming matter at an extraordinary rate—possibly **26 M⊙ per year**—allowing it to grow far more rapidly than predicted by classical accretion models.

At the same time, the quasar emits **intense X-ray radiation and powerful radio jets.** These jets carry enormous amounts of energy and can influence the surrounding galaxy by heating interstellar gas, regulating star formation, and affecting how galaxies and their central black holes evolve together.

The quasar therefore shines brightly in both X-ray and radio wavelengths, indicating the presence of both rapid accretion and energetic jet activity—a combination that challenges existing theoretical models.

Scientific Importance: The coexistence of **super-Eddington accretion and powerful jet** production is not easily explained by current theoretical models. For this reason, the newly discovered quasar provides an important laboratory for studying how supermassive black holes grew in the early universe. It may represent a previously unknown phase in the evolution of young quasars.

Discovery: This unusual quasar was identified by a research team from **Waseda University and Tohoku University, led by Sakki Obuchi**, using the **Subaru Telescope and its near-infrared spectrograph.** The discovery was reported in *The Astrophysical Journal on January 21, 2026.*

Concluding Remarks

These remarkable systems are not merely cosmic curiosities. They serve as natural laboratories for testing the limits of gravity, matter, energy, and spacetime under conditions far beyond anything possible on Earth. Some reveal how black holes grow and merge over cosmic time, while others show how they can power jets, shape galaxies, and illuminate the distant early universe. As telescopes and detection methods continue to improve, future discoveries will almost certainly uncover even stranger and more extraordinary black holes. Each new finding deepens our understanding of these mysterious.

Chapter 8

The Growth, Fate, and Significance of Black Holes

Introduction

Black holes are often introduced as the ultimate endpoints of stellar evolution. But their story does not end with formation. In reality, black holes are dynamic entities that continue to evolve, interact with their surroundings, and influence the structure of the universe on both small and cosmic scales. At the same time, processes associated with rotating black holes—such as the Penrose process and the Blandford–Znajek process—demonstrate that black holes are not merely cosmic sinks, but also potential sources of enormous energy. Through the quantum mechanical effect known as Hawking radiation, they can gradually lose mass over time, leading to a slow process of evaporation that defines their ultimate fate.

Their importance lies in their profound influence on galaxy formation, their role as natural laboratories for testing fundamental physics, and their potential to reveal new insights into energy processes in the universe.

How Large Can a Black Hole Grow: Growth Limits and Accretion

In the framework of general relativity, there is no theoretical upper limit to the mass a black hole can attain. Einstein's equations allow black holes to grow arbitrarily massive without destroying their structure. Any limits on black hole growth therefore arise not from fundamental physics, but from astrophysical processes in their surroundings.

Instability in Accretion Disk

Theoretical studies suggest that when a black hole reaches a mass of roughly **270 billion M⊙,** its accretion disk may become unstable. At such extreme masses, the disk can fragment into clumps or break apart, greatly reducing the efficiency with which the black hole can continue to grow by steadily accreting gas.

This instability does not prevent further growth altogether. While accretion may slow or even stall, black hole mergers can still increase the mass beyond this scale. Thus, the proposed threshold represents an upper limit for efficient accretion, not a fundamental upper limit on black hole growth.

Hawking Radiation

Black Hole Evaporation

For a long time, black holes were considered completely indestructible—regions from which nothing could escape, not even light, matter, or information. However, this perception changed with the groundbreaking concept of **Hawking radiation**, introduced by physicist **Stephen Hawking**. His theory suggests that black holes are not entirely black; instead, they slowly evaporate over time due to quantum effects occurring near their event horizons.

Stephen Hawking (1942–2018)

Image Credit: NASA (Public Domain)

According to **quantum mechanics**, virtual particle pairs constantly appear and vanish near the event horizon. Occasionally, one particle escapes while the other falls into the black hole. The escaping particle effectively carries away a tiny amount of the black hole's energy. Since energy and mass are equivalent, this process causes the black hole to gradually lose mass.

This gradual loss implies that a black hole can eventually shrink and disappear entirely. However, for most known black holes, this evaporation process is unimaginably slow.

For instance, a black hole with a mass equal to that of our Sun would take far longer than the current age of the universe to evaporate completely.

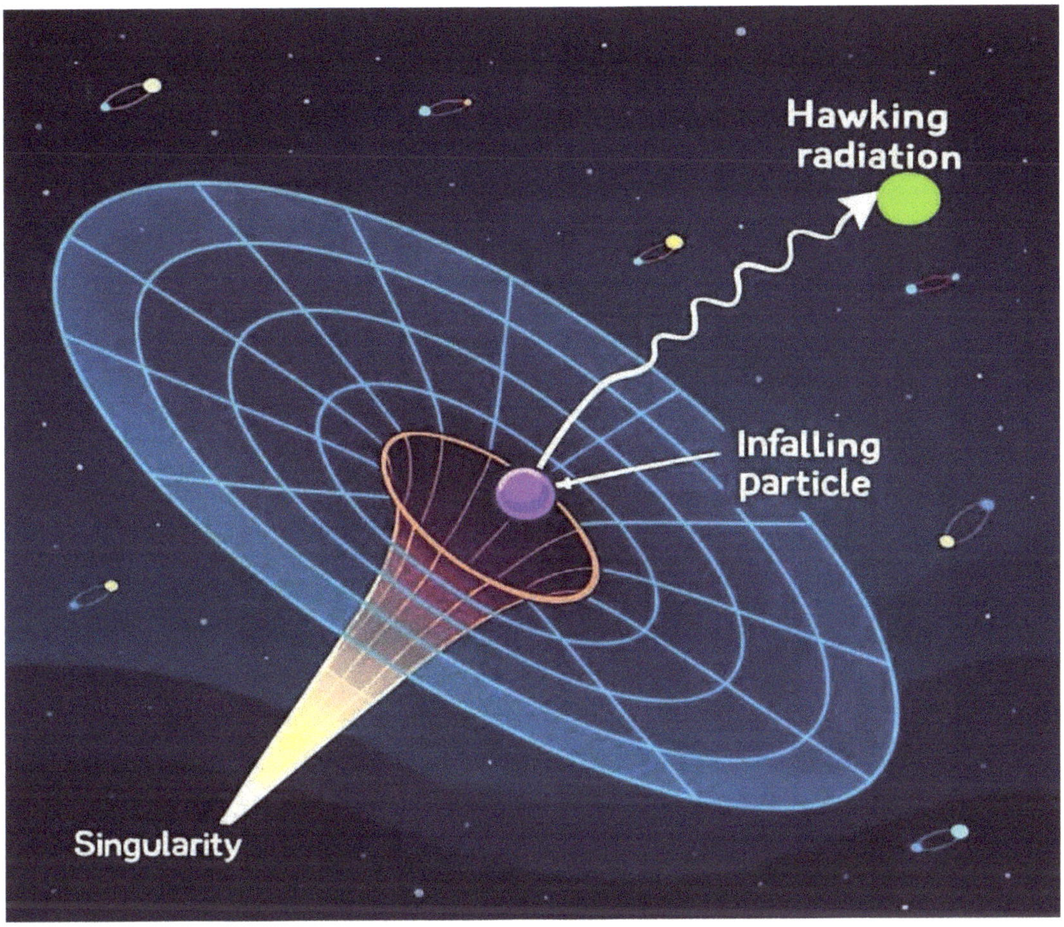

Figure 7.2: AI-generated Illustration of hawking radiation for conceptual purposes

Hawking radiation might be observable in **micro black holes**—hypothetical black holes with very small masses possibly comparable with the mass of an asteroid. Such tiny black holes would evaporate rapidly, emitting detectable bursts of radiation. Yet, no such micro black holes have been observed so far, and their existence remains purely theoretical despite ongoing research. Hawking radiation is not even detected despite several major efforts. Remains only theoretical abstracts.

1. Penrose Process – Extracting Energy from a Spinning Black Hole

In 1969, British physicist **Roger Penrose** proposed a remarkable idea that energy could, in theory, be extracted from a **rotating black hole**. Surrounding every spinning black hole is a region called the **ergosphere**, where the intense drag of spacetime forces everything to move in the direction of the black hole's rotation. Penrose suggested that if a particle enters this region and splits into two parts, one fragment could fall into the black hole with *negative energy* while the other escapes with *extra energy*—more than the original particle possessed.

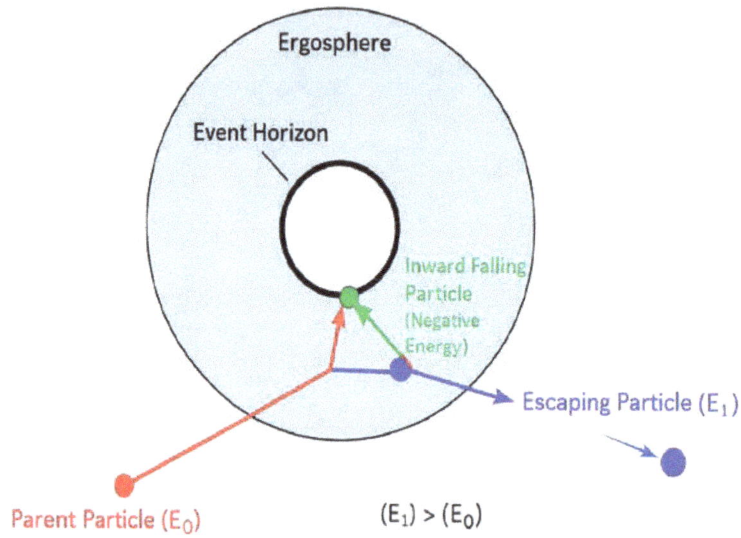

Figure 7.3: AI-generated illustration of Penrose Process for explanatory purposes

The additional energy of the escaping particle comes from the **black hole's own rotation**, effectively drawing on its **spin energy**. As this process continues, the black hole would gradually **lose angular momentum** and slow down. Ultimately, it could become a **non-rotating (static) black hole**, with all its rotational energy extracted.

Roger Penrose (1931–), English mathematical physicist and Nobel Laureate (1920),

Photograph: Via Wikimedia Commons, licensed under Creative Commons Attribution 3.0 Unported (CC BY 3.0).

Although no direct observational evidence yet confirms this process, it remains a fascinating and mathematically sound prediction of general relativity.

Differences Between Hawking Radiation and the Penrose Process

1. **Nature of the Process:** Hawking radiation is a quantum mechanical phenomenon, arising from quantum effects near the event horizon.

   The Penrose process is a classical relativistic mechanism, based on Einstein's general theory of relativity.

2. **Effect on the Black Hole:** Through Hawking radiation, a black hole loses mass and can, in theory, be completely evaporated over extremely long timescales.

   The Penrose process does not destroy a black hole; instead, it extracts rotational (spin) energy, potentially slowing a rotating black hole down to a non-rotating state.

3. **Region where the Process Occurs:** Hawking radiation originates at the event horizon of the black hole.

   The Penrose process operates within the ergosphere, a region outside the event horizon of a rotating black hole.

4. **Requirement of Black Hole Rotation:** Hawking radiation can occur for any black hole, rotating or non-rotating.

   The Penrose process requires a rotating black hole and cannot occur for non-rotating black holes.

5. **Energy Source:** Hawking radiation draws energy from quantum fluctuations of spacetime.

   The Penrose process extracts energy from the rotational energy of the black hole.

6. **Timescale and Observability:** Hawking radiation is extremely weak for astrophysical black holes and becomes significant only over timescales far exceeding the age of the universe.

   The Penrose process is theoretically more efficient but remains unverified observationally in astrophysical settings.

2. Blandford-Znajek Process

Extracting Energy from a Black Hole's Spin and Magnetic Field

The Blandford–Znajek process, proposed in **1977,** is based on general relativity combined with electromagnetism and was developed at a time when direct black hole observations were still in their early stages. Unlike Hawking radiation and the Penrose process, which remain largely theoretical or idealized, the Blandford–Znajek mechanism provides a **more astrophysically realistic description of how energy can be extracted from spinning black holes.**

In this process, the powerful magnetic fields anchored in the accretion disk directly extract the rotational energy of a spinning (Kerr) black hole and convert it into energy carried away by relativistic jets. Thus, the jets are ultimately powered by the black hole's own spin. Today, this mechanism is widely regarded as the most plausible explanation for the enormous jets observed in quasars and active galactic nuclei, and it is strongly supported by numerical simulations and indirect observational evidence.

However, the Blandford–Znajek process does not lead to the complete evaporation of a black hole. It only reduces the black hole's spin, potentially slowing a rapidly rotating black hole toward a non-rotating state. The black hole itself remains intact, with its mass largely preserved. Thus, while this mechanism may be more relevant in real astrophysical environments than the Hawking or Penrose processes, it describes energy extraction rather than black hole evaporation or disappearance.

Mass–Energy Loss in Black Holes

Black holes can lose energy and, in certain situations, effectively reduce their mass through different physical processes but still these processes are hypothetical and have no confirmatory observations.

In contrast, black hole mergers provide direct observational evidence of mass-energy loss. During such events, a fraction of the system's total mass is converted into Gravitational waves, as confirmed by detections from observatories like LIGO. This demonstrates that black holes can lose mass-energy during dynamic interactions, although this occurs specifically during merger events rather than as a continuous process.

Significance of Black Holes

Black holes—often described as the "**monsters**" of the cosmos—are, in fact, remarkably beneficial to our understanding of the universe. Rather than being merely destructive forces, they serve as **natural laboratories** that allow scientists to study the most extreme conditions in nature. They are fascinating, mysterious, and, in many ways, friendly to science. Black holes are the cosmos's deepest devourers, yet their gravity can turn destruction into light so intense that an entire galaxy glows around them. They are also the sources of **gravitational waves**—ripples in spacetime that help us to explore the earliest moments of the universe.

Significance of Black Holes in the Galaxy

Almost every massive galaxy in the universe harbors a supermassive black hole (SMBH) at its center. Far from being destructive entities, these cosmic giants play a crucial role in shaping and sustaining their host galaxies. Black holes also contribute to the long-term **stability and evolution** of galaxies. Their immense gravity helps maintain the orbital harmony of billions of stars, preventing large-scale chaos or stellar collisions. In this sense, they act as **invisible anchors**, holding galaxies together over cosmic timescales.

Our Solar System, being part of the Milky Way, indirectly benefits from the stability ensured by SagittariusA*. The calm, dormant state of our galactic black hole may even have been favorable for the development of life on Earth, allowing our region of the galaxy to remain stable and habitable for billions of years.

Thus, black holes are not merely cosmic destroyers, as once imagined, but **essential architects of galaxies**—governing their formation, evolution, and stability, and ultimately helping to create the cosmic conditions that made life possible on our planet.

Natural Particle Accelerator

Black holes can also act as natural particle accelerators. Near supermassive black holes, powerful magnetic fields and rapid rotation can accelerate particles to nearly the speed of light, with some of this high-energy matter being launched outward in relativistic jets. No human-made accelerator can match these extreme conditions.

Quasars: Windows into the Early Cosmos

Quasars are invaluable to humanity because they allow us to see the universe as it once was. Acting as natural cosmic time machines, their ancient light travels billions of years before reaching Earth, revealing conditions in the young universe shortly after the Big Bang. By studying quasar spectra and variability, astronomers can trace how matter, radiation, and cosmic structures evolved across precise intervals of cosmic time.

Black Holes: The Friendly Giants of the Cosmos

Black holes are, in many ways, the most cooperative objects in the universe. Though often portrayed as devourers, they frequently coexist peacefully with stars, neutron stars, and even other black holes, forming remarkable binary and multiple systems. Their gravitational bonds—immensely powerful yet elegantly balanced—are among the most fascinating phenomena in the cosmos.

Despite the hostile environments that surround them, black holes can form **gravitational partnerships** that astonish astronomers. For example, in the **V404 Cygni** system, a distant star orbits a binary composed of a black hole and a companion star, creating a dramatic interplay of gravity and light. Similarly, in the **D9 system**, a binary star system orbits internally while simultaneously revolving around Sagittarius A*, remaining stable within its strong gravitational field despite the harsh conditions.

Such systems remain puzzles for scientists, revealing the complex harmony that black holes can maintain with their cosmic companions.

These examples remind us that black holes are not merely destroyers—they are **balancers of cosmic order**. Their immense gravity orchestrates the motion of stars and gas, sustaining the galactic ecosystem. Astronomers continue to search for more such extraordinary systems, where black holes act not as foes but as **guardians of stability** in the grand symphony of the universe.

The Benevolent Side of Black Holes

While black holes appear friendly to their cosmic neighbors, they are not kind to intruders. Any object or spacecraft venturing too close—crossing the **event horizon**, the ultimate point of no return—would be lost forever to the depths of spacetime. Their friendship ends precisely where their boundary begins.

Yet, from a safe distance, black holes remain among humanity's greatest **cosmic allies**—guardians of galactic structure, companions of stars, and keys to understanding the fundamental laws of nature.

The study of black holes provides profound insights into **fundamental physics, cosmology, and spacetime**. They allow us to test Einstein's **general relativity** and probe the limits of our understanding of gravity, mass, and time. Research into black holes has also inspired new technologies, especially in **detection methods**, **computational modeling**, and **data analysis**, with applications far beyond astronomy.

Despite decades of study, the true nature of the **singularity**—the central point where physics as we know it breaks down—remains elusive. Scientists continue to develop new theories and models to uncover its secrets.

Black Holes: The Most Challenging Cosmic Giants

Black holes stand among the most perplexing and intellectually demanding objects in the universe. They have repeatedly forced astronomers and physicists to revise long-held theories, revealing exceptions to almost every rule once thought secure. For decades, stellar evolution models stated that massive stars must explode as supernovae to form black holes, yet discoveries now show black holes that appear to have formed without any detectable explosion, suggesting entirely new formation pathways. Similarly, jet physics once assumed that only rapidly spinning black holes could launch powerful relativistic jets, but observations have uncovered **slowest-spinning black holes still producing narrow, energetic jets**, challenging our understanding of how magnetic fields extract energy from black holes.

Even the limits designed to describe black hole feeding behaviour have been broken. The celebrated **Eddington limit**—long considered the maximum sustainable feeding rate—

has been exceeded by several black holes in the early universe, indicating that extreme environments can allow matter to fall inward far faster than theory predicts. These surprising findings show that black holes are not just exotic gravitational traps but **active laboratories of extreme physics**, where gravity, radiation, magnetism, and spacetime behave in ways that remain beyond the reach of current theory.

In this sense, black holes are not only cosmic giants that challenge every theory—they are also **essential tools for unlocking the deepest laws of nature**.

The number of scientists and astronomers studying black holes is increasing rapidly, reflecting the growing interest and momentum in this field. Many key advances in black hole physics have been recognized with Nobel Prizes, reflecting the profound importance of black holes in modern astrophysics.

Black Holes in Popular Culture and Media

The mystery and power of black holes have captivated not only scientists but also filmmakers and storytellers. Their portrayal in science-based films such as **Interstellar, Event Horizon, Star Trek, The Black Hole, and High Life (2018)** has inspired audiences by blending imagination with elements of scientific realism. Many of these productions became commercially successful and widely discussed, reflecting the public's deep fascination with black holes. Their immense gravity, strange effects on time and space, and air of cosmic mystery continue to intrigue viewers, showing how strongly these enigmatic objects capture human curiosity and imagination.

In recent years, black holes have become vivid subjects of **visual science communication**. Across digital media and educational platforms, scientists share stunning simulations and documentaries that reveal their hidden mechanisms. Viewers can now witness **spaghettification**, where matter is stretched by gravity's immense pull, experience **time dilation**, where time slows near the event horizon and explore phenomena like **tidal disruption events**, **binary mergers**, and **relativistic spin**, all through breathtaking visualizations.

These explorations are far more than entertainment—they are **windows into cosmic truth**. They bridge art and science, ignite curiosity in young minds, and renew wonder in all who gaze at the night sky. Once feared as devourers of light, black holes now **illuminate our understanding** of physics, relativity, and the unity of the cosmos. In learning about what they consume, we discover how much they give back to human knowledge.

Finally, one of the most imaginative ideas is the concept of a **black hole starship**—a theoretical spacecraft powered by black hole energy, capable of interstellar travel. Though purely speculative, it reflects an important truth: ***"Imagination is the first and finest step toward transforming science fiction into scientific discovery."***

The Continuing Journey

The story of black holes began not with telescopes, but with theory. In **1784**, the English natural philosopher **John Michell** proposed the idea of a **"dark star"** so massive that even light could not escape its gravity—a remarkably prescient concept for its time. The idea was echoed soon after by **Pierre-Simon Laplace**, but it faded into obscurity for more than a century.

It was **Albert Einstein's theory of general relativity**, published in 1915, that rekindled interest in the possibility of such extreme gravitational objects. His field equations quietly contained the blueprint for regions of space where gravity could become infinitely strong. Soon after, in the same year, **Karl Schwarzschild** discovered the first exact solution to Einstein's equations, describing how space and time would behave around a point of infinite density — what we now recognize as a **black hole**. Decades later, **Subrahmanyan Chandrasekhar** provided the crucial theoretical bridge between stellar physics and relativity. In the early 1930s, his work on the **mass limit of white dwarfs** showed that beyond a certain point, no known force could stop a massive star from collapsing under its own gravity. This insight offered the first realistic pathway for black hole formation in nature. Over the following decades, the theoretical framework grew stronger, until in **1971**, the discovery of **Cygnus X-1** offered the first convincing observational evidence of a black hole within our own galaxy — transforming a once-theoretical idea into an astrophysical reality.

Following the discovery of Cygnus X-1, many other black hole systems have been identified. From hyperluminous quasars such as TON 618, shining with the light of trillions of Suns, to ravenously feeding black holes, giant relativistic jet systems like Porphyrion, and candidate close supermassive black hole pairs such as Markarian 501, each new discovery expands the boundaries of our knowledge.

Detection of Gravitational Waves

A major leap came with the first detection of gravitational waves by LIGO during the GW150914 event, providing direct evidence of black hole mergers. In these cataclysmic collisions, part of the black holes' mass is released as ripples in spacetime itself. Later events, including GW190521, revealed even more massive mergers. Most profoundly, gravitational-wave astronomy has given humanity a new way to observe the universe — not by light alone, but by listening to the faint echoes of cosmic collisions across space and time.

These discoveries highlight that black holes are not only common but also dynamically active participants in shaping galaxies and the large-scale structure of the universe.

A Note on Scientific Progress

The information presented in this book reflects the current understanding of black holes up to the time of writing. It is important to remember that this knowledge will continue to evolve. New discoveries will emerge, while earlier interpretations may be refined, revised, or replaced. Such evolution is a natural and essential part of scientific progress.

Possible Future Discoveries of Black Holes

The story of black holes is far from complete. Future telescopes and gravitational-wave observatories may reveal hidden populations of intermediate-mass black holes, confirm more binary mergers, and even capture detailed images of event horizons in distant galaxies. Future discoveries are expected to reveal many more supermassive black hole binaries and rogue black holes. Scientists estimate that thousands of rogue black holes may exist throughout the universe, but so far only a very small number, not more than five or six, have been discovered and confirmed. As observational techniques improve, many more such hidden black holes are likely to be detected. Future observatories such as the **Einstein Telescope** and **Laser Interferometer Space Antenna(LISA)** will greatly expand our ability to detect gravitational waves. While the Einstein Telescope will probe fainter and more distant mergers from deep underground on Earth, LISA will detect lower-frequency waves from supermassive black hole mergers in space, opening an even broader window onto the hidden universe.

Advances in technology and theory could uncover how black holes formed in the early universe and whether they hold clues to quantum gravity itself. The exploration of these cosmic enigmas will continue—pushing the boundaries of human curiosity and imagination.

The study of black holes remains a dynamic and ever-evolving field—one that continually pushes the boundaries of human knowledge. As technology advances and observations deepen, each new discovery brings us a little closer to unlocking the ultimate mysteries of gravity, matter, and the universe itself.

"The journey that began with a single question centuries ago is, in truth, the story of humanity itself—our endless desire to understand the darkness, and to find light even within it."

Appendix: A

Glossary of Scientific Terms

Active Galactic Nuclei (AGN): An active galactic nucleus (AGN) refers to the active phase of that object when it is accreting matter. An SMBH can be inactive and quiet, but an AGN is always active and luminous.

BL Lacertae–Type Active Galactic Nucleus (BL Lac object): It is a highly energetic galaxy whose central supermassive black hole produces a powerful jet of particles directed almost exactly toward Earth.

Comoving Distance: The distance between two objects measured in a way that ignores the expansion of the universe, so it remains constant over time if the objects are moving with the cosmic expansion.

Dark Energy: A mysterious form of energy that permeates all of space and is responsible for the accelerated expansion of the universe. It does not emit, absorb, or reflect light and is detected only through its gravitational effects on cosmic scales. Dark energy is estimated to constitute about 68–70% of the total energy content of the universe and is commonly modeled as a cosmological constant (Λ) in modern cosmology.

Dynamical Modeling: Dynamical modeling is a method used to measure the mass of a black hole by studying the motion of nearby stars or gas under its gravitational influence.

Electron Degeneracy Pressure: Electron degeneracy pressure is a quantum mechanical effect that prevents matter from being compressed beyond a certain limit. Electrons cannot all occupy the same state, so when matter is squeezed very tightly, the electrons push back strongly. This pressure supports objects like white dwarf stars against gravitational collapse.

Extragalactic Objects: The celestial objects that are located in other galaxies, not in our Milky Way.

Globular Cluster: A dense, roughly spherical collection of hundreds of thousands to millions of old stars gravitationally bound together and orbiting the outskirts.

Gravitational Waves: Gravitational waves are disturbances in spacetime generated by accelerating masses, analogous to how accelerating charges produce electromagnetic waves. They travel at the speed of light and produce minute changes in distances as they pass. Their wavelengths can range from astronomical scales to thousands of kilometers depending on frequency. Because gravity is much weaker than electromagnetism, gravitational waves interact only very weakly with matter.

Joint Global Effort: Together with the LIGO detectors in the United States, Virgo and KAGRA form the international LVK gravitational-wave observatory network, enabling precise detection and source localization of gravitational waves from black hole mergers, neutron star collisions, and other extreme cosmic events.

KAGRA: KAGRA is Japan's laser interferometer gravitational-wave detector located underground in the Kamioka mine. It is developed by the Institute for Cosmic Ray Research (ICRR) at the University of Tokyo, with contributions from the National Astronomical Observatory of Japan (NAOJ) and High Energy Accelerator Research Organization (KEK). KAGRA is the first underground gravitational-wave detector and uses cryogenic mirrors to reduce noise. The KAGRA Collaboration includes over 400 scientists from about 128 institutes spanning 17 countries and regions who contribute to detector development, operation, and science as part of the LVK network.

Lense–Thirring Effect: Also known as frame dragging, it is a phenomenon in which a rotating massive object—such as a spinning black hole or even the Earth—drags the surrounding spacetime along with it.

M-Type Red Dwarf: A small, cool, and faint star of spectral type M that produces energy through nuclear fusion, appears reddish in color, and can live for extremely long periods of time.

Magnetorotational Instability (MRI): A process in rotating, magnetized disks in which angular momentum is transferred outward, allowing matter to lose momentum and spiral inward toward the central object.

Main Sequence: The normal, stable life of a star when it truly becomes a star and shines steadily.

Microquasar: A microquasar is a stellar-mass analogue of a quasar. While quasars are powered by supermassive black holes at the centers of distant galaxies, microquasars are powered by stellar-mass black holes in binary systems within our own galaxy. Both exhibit accretion disks and relativistic jets governed by the same physical processes, differing mainly in scale and mass.

Neutron Degeneracy Pressure: A quantum pressure that prevents neutrons from being squeezed closer together, supporting a neutron star against gravitational collapse.

Nuclear Fusion: Nuclear fusion is the process in which light atomic nuclei combine to form a heavier nucleus, releasing a large amount of energy. It is the reaction that powers stars, including our Sun, by converting hydrogen into helium and producing the light and heat we receive.

O-Type Star: An O-type star is an extremely hot, massive, and very luminous young star. It burns its nuclear fuel at a tremendous rate, shines with a bluish color, and strongly influences its surroundings with intense radiation and stellar winds. O-type stars live short lives and often end as supernovae, leaving behind neutron stars or black holes.

Pair Instability: Pair instability is a process that occurs in extremely massive stars when their cores become so hot that high-energy gamma rays turn into pairs of electrons and positrons. This sudden creation of particle pairs reduces the pressure that was supporting the star against gravity. As a result, the core contracts rapidly, triggering violent nuclear reactions that can cause the star to partially or completely explode.

R-Process (Rapid Neutron Capture Process): A nucleosynthesis process in which atomic nuclei rapidly capture neutrons in extremely neutron-rich environments before radioactive decay can occur. It is responsible for producing many of the heaviest elements in the Universe, such as gold, platinum, and uranium, particularly during neutron-star mergers and some supernova explosions.

Red Giant Star: A red giant is a star that has exhausted the hydrogen fuel in its core and has begun burning hydrogen in a surrounding shell. As a result, the star expands enormously and becomes much larger and brighter, but cooler at its surface, giving it a reddish color. Red giants represent a late stage in the life of stars like our Sun.

RBH1 (Runaway Black Hole 1): A black hole that has been ejected from its original location—such as a galaxy or star system—due to gravitational interactions, moving through space at high speed.

Seyfert Galaxy: A Seyfert galaxy is a type of galaxy with a bright, active nucleus powered by a supermassive black hole at its center. These galaxies emit strong radiation, especially in ultraviolet and X-ray wavelengths, due to hot gas accreting onto the black hole. Seyfert galaxies are among the nearest and most common active galaxies, making them important for studying black hole activity.

Spectrophotometric Reverberation Mapping: Spectrophotometric reverberation mapping can be used to measure the mass of a black hole. It works by observing the time delay between changes in the central light source and the response of surrounding gas clouds, which gives their distance from the center. At the same time, spectroscopy shows how fast this gas is moving. Combining the size of this region with the speed of the gas allows scientists to calculate the mass of the black hole, since stronger gravity means faster motion.

Tidal Disruption Event (TDE): A tidal disruption event occurs when a star passes too close to a black hole and is torn apart by its strong gravity. The stellar debris forms a temporary accretion disk around the black hole, producing a sudden and bright flare of radiation that can be observed across many wavelengths.

Torus: In black hole physics especially for active galactic nuclei, (AGN), a torus is a thick, donut-shaped ring of gas and dust surrounding the central black hole and its accretion disk.

Vacuum Energy: Vacuum energy is the energy that exists even in completely empty space. According to quantum physics, space is never truly "empty." Instead, it is filled with tiny, constantly fluctuating fields that briefly create and destroy particles. These fluctuations carry energy, called vacuum energy.

Virtual Earth-Size Telescope: A Virtual Earth-Size Telescope is not a single physical instrument. It is created by linking many radio telescopes spread across the Earth and making them work together as one. This technique is called Very Long Baseline Interferometry (VLBI).

By combining signals from telescopes on different continents, scientists achieve the same resolving power as a telescope as large as the distance between them—almost the size of the Earth itself. While a traditional telescope is a single physical instrument located at one place. Its size and resolving power are limited by the diameter of its main mirror or dish. Larger mirrors give sharper images, but building extremely large single telescopes is technologically difficult and expensive.

Virgo Detector: The Virgo detector is a large laser-interferometer gravitational-wave observatory located in Italy and managed by the European Gravitational Observatory (EGO), a consortium formed by French and Italian research institutions. Virgo is part of the global LVK network (LIGO-Virgo-KAGRA) that detects and analyzes gravitational waves from merging compact objects. The Virgo Collaboration consists of roughly 940 scientists from about 165 research institutions in over 20 countries, working on everything from detector commissioning and operation to data analysis and astrophysical interpretation.

Wobble: A gentle, repeating shift in a star's position caused by the gravity of an unseen companion.

Appendix: B

Understanding Astronomical Names

1) **3C 48:** 3C refers to the Third Cambridge Catalogue of Radio Sources; 48 is the object's entry number.

2) **3C 186:** 3C refers to the Third Cambridge Catalogue of Radio Sources; 186 is the entry number.

3) **3C 273:** 3C refers to the Third Cambridge Catalogue of Radio Sources; 273 is the entry number.

4) **4C+37 11:** 4C refers to the Fourth Cambridge Catalogue of Radio Sources; +37 indicates the declination zone; 11 is the source number within that zone.

5) **A0620–00:** A refers to the Ariel 5 satellite; 0620 indicates right ascension (06h 20m);–00 indicates declination near 0°.

6) **AT2024tvd:** AT stands for Astronomical Transient; 2024 is the year of discovery; tvd is a sequential designation code.

7) **CAPERS-LRD-z9:** CAPERS refers to the survey name; LRD means "Luminous Red Disk"; z9 indicates a redshift of about 9.

8) **GW** (Gravitational Wave Events)**:** "GW" denotes gravitational wave detections; the numbers indicate the date of detection (e. g. , GW150914 → 14 September 2015).

9) **G3425:** G refers to the Gaia catalogue; 3425 is the entry number.

10) **Gaia BH1:** Gaia refers to the ESA Gaia mission; BH1 indicates the first black hole discovered in that dataset.

11) **GRO J0422+32:** GRO stands for Gamma-Ray Observatory; J indicates J2000 coordinates; 0422 gives right ascension; +32 gives declination.

12) **GRS 1915+105:** GRS stands for Gamma Ray Source; 1915+105 gives sky coordinates (RA 19h 15m, Dec +10° 5′).

13) **HDE 226868:** HDE refers to the Henry Draper Extension catalogue; 226868 is the object number.

14) **IC 1101:** IC refers to the Index Catalogue; 1101 is the entry number.

15) **J0529–4351:** J indicates J2000 coordinates; 0529 is right ascension; –4351 is declination.

16) **J1218+2953 / J1219+1035:** Coordinate-based designations using the J2000 system (right ascension and declination).

17) **J152932. 16+601534. 4:** High-precision J2000 coordinates (RA 15h 29m 32. 16s, Dec +60° 15′ 34. 4″).

18) **J1610+1811:** J indicates J2000 coordinates; 1610 is right ascension; +1811 is declination.

19) **LID 568:** An internal source identification number used in a specific survey; 568 is the entry number. UHZ1

20) **LMC X-1:** LMC refers to the Large Magellanic Cloud; X-1 indicates the first identified X-ray source.

21) **M31-2014-DS1:** M31 refers to the Andromeda Galaxy; 2014 is the year; DS1 is an internal designation ("Dusty Star 1").

22) **M33 X-7:** M33 refers to the Triangulum Galaxy; X-7 indicates the seventh X-ray source.

23) **M87*:** M refers to the Messier Catalogue; 87 is the entry number; * indicates the central supermassive black hole.

24) **NGC 1365: NGC** refers to the New General Catalogue; 1365 is the entry number.

25) **NGC 4395:** NGC refers to the New General Catalogue; 4395 is the entry number.

26) **NGC 4889:** NGC refers to the New General Catalogue; 4th is the entry number.

27) **NGC 6240:** NGC refers to the New General Catalogue; 6240 is the entry number.

28) **NGC 6946-BH1:** NGC refers to the New General Catalogue; 6946 is the galaxy; BH1 indicates Black Hole candidate 1.

29) **NGC 7727:** NGC refers to the New General Catalogue; 7727 is the entry number.

30) **OB110462:** OGLE-based designation; 11 indicates the year 2011; 0462 is the event number.

31) **OGLE-2003-BLG-235Lb:** The first planet ("b") discovered around a lens star ("L") in the 235th microlensing event observed in 2003 by the Optical Gravitational Lensing Experiment, toward the Galactic.

32) **OJ7:** Originates from the Ohio Sky Survey;287 is the catalogue number (OJ is a combined survey code).

33) **PA-99-N2:** PA refers to the POINT-AGAPE survey; 99 is the year 1999; N2 is the event number.

34) **QSO J0529–4351:** QSO means Quasi-Stellar Object; J indicates J2000 coordinates; numbers give sky position.

35) **TON 618:** TON refers to the Tonantzintla Observatory catalogue; 618 is the entry number.

36) **UHZ1:** Stands for Ultra-High-Redshift source 1.

37) **V404 Cygni:** The 404th variable star discovered in the constellation Cygnus.

38) **W49B:** Refers to the "B" component of the 49th radio source in the Westerhout catalogue.

39) **XTE J1650-500:** stands for X-ray Timing Explorer source at right ascension 16h 50m, declination –50°.

Appendix: C

Vocabulary

abruptly——-----------------sudden a unexpected

abstract——-----------------existing only as idea

abyssis——------------------A very deep bottomless hole or space

anchored——----------------to fix firmly

anomolies——---------------somethinthat deviate from normal

avid——----------------------eager

bulge——---------------------to be full of something

cataclysmic——-------------violently destructive

catastrophes——-----------a sudden disaster

chaotic——------------------unsettled, confused

coherent——----------------maintaining fixed phase

colossal——-----------------extremely large

constraining——------------placing limits on the possible values

counterintuitive——--------contray to common-sense

cumulative——--------------increasing steadily

culminative——-------------reaching a climax or highest point

dazzling——-----------------shining intensely

devour——-------------------eat quickly

elusive——-------------------not easy to catch

episodic——-----------------contaning a series of event

epoch——---------------------age, period or time

exotic——---------------------something unusual

formidable——--------------inspiring fear

flung—-------------------to throw with lot of for

enigmatic—--------------mysterious

eerie—--------------------strange and frightening

extragalactic—-----------other than Milky Way

fragile—------------------delicate

furious—------------------very angry

ground breaking—--------finding out new thing using new methods

hierarchical—------------something arranged by levels of importance

inadvertent—-------------unintentional

Inherently—--------------exists as a natural

ingenuity —--------------simplicity

impeding—---------------to make it difficult for somebody

intrigue —----------------interesting

inaccessible—------------difficult to reach

kin------------------------one's family and relations

minimal—----------------very small in amount, size or level

orchestrates—------------plan for arrangements

paradigm—---------------set of ideas

perpetually—-------------continually

plausible—---------------appreciable

plunge—------------------rapid fall

precise—------------------exact, good

precursors—--------------pioneers

prescient—---------------knowing about event before it takes place

profound—---------------very great

progenitor—--------------thing from the past

quiescent—---------------calm, inactive

radically—----------------completely or extremely

ravenously—-------------excessively greedy

reinforce—---------------to make stronger

rekindled—-------------to make something active again

silhouette—-------------something seen against light background

spatial—----------------related to space

speculative—----------based on guessing

sporadic—--------------not happening regularly

Stalled ------------------to avoid doing some thing

stupendous—----------very large or impressive

subtle—-----------------difficult to detect or analyse

synchronous—---------existing or occuring at the same time

swamp—----------------to cover

tangible—--------------robust

tapestry —--------------thread to a complete unique structure

tranquil—---------------quiet or peaceful

unaccountable—-------difficult to explain

unchartered—----------not mapped

unprecedented—-------never before

unpinned —------------removed

unravel—---------------something become clear

untethered—-----------unbound,

veil —-------------------mask

vicinity—---------------near about

vindicated—-----------clear from censure or doubt

waltz—-----------------an elegant dance with partner

wobble—---------------to move from side to side that shows poor balance

References

Chapter 1 – Discovery and Detection of Black Holes

1. Michell, J. (1784).

On the Means of Discovering the Distance, Magnitude, &c. of the Fixed Stars.

Philosophical Transactions of the Royal Society of London, **74**, 35–57.

2. Laplace, P. S. (1796).

Exposition du Système du Monde. Paris: Cercle-Social.

3. Einstein, A. (1916).

The Foundation of the General Theory of Relativity.

Annalen der Physik, **49,** 769–822.

4. Schwarzschild, K. (1916).

On the Gravitational Field of a Point Mass According to Einstein's Theory.

Sitzungsberichte der Königlich Preußischen Akademie der Wissenschaften, **189–**196.

5. Carroll, S. (2019).

Spacetime and Geometry: An Introduction to General Relativity.

Cambridge University Press.

6. Hartle, J. B. (2003).

Gravity: An Introduction to Einstein's General Relativity.

Addison-Wesley.

7. Schutz, B. F. (2009).

A First Course in General Relativity.

Cambridge University Press.

8. Bolton, C. T. (1972).

Identification of Cygnus X-1 with HDE 226868.

Nature, **235,** 271–273.

9. Orosz, J. A., McClintock, J. E., Aufdenberg, J. P., et al. (2011).

Modeling the Optical and X-Ray Observations of Cygnus X-1: Determination of the Mass of the Black Hole and Its Companion.

The Astrophysical Journal, **742**, 84.

10. Webster, B. L., & Murdin, P. (1972).

Cygnus X-1: A Spectroscopic Binary with a Heavy Companion.

Nature, **235,** 37–38.

11. Frank, J., King, A., & Raine, D. (2002).

Accretion Power in Astrophysics (3rd ed.).

Cambridge University Press.

12. Krolik, J. H. (1999).

Active Galactic Nuclei: From the Central Black Hole to the Galactic Environment.

Princeton University Press.

13. Bambi, C. (Ed.). (2017).

The Astrophysics of Black Holes.

Springer.

14. Abbott, B. P., Abbott, R., Abbott, T. D., et al. (2016).

Observation of Gravitational Waves from a Binary Black Hole Merger.

Physical Review Letters, **116**, 061102.

15. Event Horizon Telescope Collaboration (2019).

First M87 Event Horizon Telescope Results. The Shadow of the Supermassive Black Hole.

The Astrophysical Journal Letters, **875**, L1.

16. Sedda, M.A. (Ed.). (2023).

Black Holes in the Era of Gravitational-Wave Astronomy.

Elsevier.

Chapter 2 – The Concept of Black Holes and Their Formation

1. Einstein, A. (1916).

The Foundation of the General Theory of Relativity.

Annalen der Physik, **49,** 769–822.

2. Chandrasekhar, S. (1931).

The Maximum Mass of Ideal White Dwarfs.

The Astrophysical Journal**, 74,** 81–82.

3. Oppenheimer, J. R., & Snyder, H. (1939).

On Continued Gravitational Contraction.

Physical Review, **56,** 455–459.

4. Penrose, R. (1965).

Gravitational Collapse and Space-Time Singularities.

Physical Review Letters, **14**, 57–59.

5. Misner, C. W., Thorne, K.S., & Wheeler, J.A. (1973).

Gravitation.

W.H. Freeman.

6. Thorne, K.S., Price, R. H., & Macdonald, D. A. (1986).

Black Holes: The Membrane Paradigm.

Yale University Press.

Chapter 3 – Types and Classification of Black Holes

1. Shapiro, S.L., & Teukolsky, S. A. (1983).
Black Holes, White Dwarfs, and Neutron Stars: The Physics of Compact Objects.
Wiley.

2. Remillard, R. A., & McClintock, J. E. (2006).
X-Ray Properties of Black Hole Binaries.
Annual Review of Astronomy and Astrophysics, **44,** 49–92.

3. Casares, J., & Jonker, P.G. (2014).
Mass Measurements of Stellar and Intermediate-Mass Black Holes.
Space Science Reviews, **183,** 223–252.

4. Lin, D., et al. (2018).
An Intermediate-Mass Black Hole Candidate in the Galaxy NGC 4395.
Nature Astronomy, 2, **656**–662.

5. Volonteri, M. (2010).
Formation of Supermassive Black Holes.
Astronomy and Astrophysics Review, **18,** 279–315.

6. Ghez, A. M., et al. (2008).
Measuring Distance and Properties of Sgr A*: The Supermassive Black Hole at the Galactic Center.
The Astrophysical Journal, **689,** 1044–1062.

7. Gebhardt, K., et al. (2011).
The Black Hole Mass in M87 from Gemini/NIFS Adaptive Optics Observations.
The Astrophysical Journal, **729,** 119.

8. Event Horizon Telescope Collaboration (2019).
First M87 Event Horizon Telescope Results. I. The Shadow of the Supermassive Black Hole.
The Astrophysical Journal Letters, **875**, L1.

9. Event Horizon Telescope Collaboration (2022).
First Sagittarius A* Event Horizon Telescope Results. I. The Shadow of the Supermassive Black Hole in the Center of the Milky Way.
The Astrophysical Journal Letters, **930,** L12.

Chapter 4 – Binary Systems Containing Black Holes

Stellar-Mass Black Hole–Star Binary Systems

1. Bolton, C. T. (1972).

Identification of Cygnus X-1 with HDE 226868.

Nature, **235,** 271–273.

2. Webster, B. L., & Murdin, P. (1972).

Cygnus X-1: A Spectroscopic Binary with a Heavy Companion.

Nature, **235,** 37–38.

3. Orosz, J. A., et al. (2007).

A 15. 65-Solar-Mass Black Hole in an Eclipsing Binary in the Nearby Galaxy M33.

Nature, **449,** 872–875.

4. Gou, L. , McClintock, J. E., Liu, J., et al. (2009).

The Spin of the Black Hole in LMC X-1.

The Astrophysical Journal, **701,** 1076–1090.

5. Wang, S., et al. (2024).

A Low-Mass Black Hole in a Binary System with a Red Giant Revealed by Gaia and LAMOST.

Nature Astronomy.

6. Remillard, R. A., & McClintock, J.E. (2006).

X-Ray Properties of Black Hole Binaries.

Annual Review of Astronomy and Astrophysics, **44,** 49–92.

Gravitational-Wave Binary Black Hole and Neutron-Star–Black-Hole Mergers

7. Abbott, B. P., Abbott, R., Abbott, T. D., et al. (2016).

Observation of Gravitational Waves from a Binary Black Hole Merger.

Physical Review Letters, **116,** 061102.

(Event GW150914)

8. Abbott, B.P., et al. (2016).

GW151226: Observation of Gravitational Waves from a 22-Solar-Mass Binary Black Hole Coalescence.

Physical Review Letters, **116,** 241103.

9. Abbott, R., et al. (2021).

Observation of Gravitational Waves from Two Neutron Star–Black Hole Coalescences.

The Astrophysical Journal Letters, **915,** L5.

(events GW200105 and GW200115)

10. Abbott, R., et al. (2023).

GWTC-3: Compact Binary Coalescences Observed by LIGO and Virgo.

Physical Review X, **13,** 041039.

11. Abac, A. G., et al. (2025).

GW231123: A Binary Black Hole Merger with Total Mass 190–265 M⊙.

Supermassive Black Hole Binary Systems

12. Sillanpää, A., Haarala, S., Valtonen, M.J., Sundelius, B., & Byrd, G. G. (1988).

OJ 287: A Binary Black Hole Model for the Periodic Optical Outbursts.

The Astrophysical Journal, **325,** 628–634.

13. Valtonen, M. J., Lehto, H. J., Nilsson, K., et al. (2006).

The 2005 November Outburst in OJ 287 and the Binary Black Hole Model.

The Astrophysical Journal, **646**, 36–51.

14. Valtonen, M.J., Mikkola, S., Merritt, D., Gopakumar, A., et al. (2010).

Measuring the Spin of the Primary Black Hole in OJ 287.

The Astrophysical Journal, **709,** 725–732.

15. Titarchuk, L., Liu, X., & Song, H. (2023).

A New Model for the Periodic Outbursts of the BL Lac Object OJ 287.

Astrophysics and Space Science.

16. Komossa, S., Zhou, H., & Lu, H. (2003).

A Recoiling Supermassive Binary Black Hole in the Radio Galaxy 4C+37. 11.

The Astrophysical Journal Letters, **582,** L15.

17. Volonteri, M. (2010).

Formation of Supermassive Black Holes.

Astronomy and Astrophysics Review, **18,** 279–315.

Chapter 5 – Extreme Cosmic Systems, Objects and Limits

Extreme Binary and Triple Systems

1. **Burdge, K.B., El-Badry, K., Kara, E., Canizares, C., Vanderburg, A., et al. (2024).**

The Black Hole Low-Mass X-Ray Binary V404 Cygni Is Part of a Wide Hierarchical Triple.

NATURE, **635,** 316–320.

2. **Peißker, F., et al. (2024).**

Binary Stars Orbiting the Supermassive Black Hole at the Galactic Center.

Nature Communications.

3. **Clowes, R.G., Harris, K.A., Raghunathan, S., et al. (2013).**

A Structure in the Early Universe at z ≈ 1. 3 that Exceeds the Homogeneity Scale of the R-W Concordance Cosmology.

Monthly Notices of the Royal Astronomical Society, **429,** 2910–2916.

(Large Quasar Group)

4. **Kollatschny, W., Weilbacher, P. M., Ochmann, M. W., Chelouche, D., Monreal-Ibero, A., Bacon, R., & Contini, T. (2020).**

NGC 6240: A Triple Nucleus System in the Final Stage of Merging.

Astronomy & Astrophysics, **633**, A79.

Quasars, Blazars and Large Quasar Structures

5. **Schmidt, M. (1963).**

3C 273: A Star-Like Object with Large Redshift.

Nature, **197,** 1040.

6. **Shen, Y., Richards, G.T., Strauss, M. A., et al. (2011).**

A Catalog of Quasar Properties from Sloan Digital Sky Survey Data Release 7.

The Astrophysical Journal Supplement Series, **194,** 45.

7. **Clowes, R.G., Harris, K.A., Raghunathan, S., et al. (2013).**

A Structure in the Early Universe at z ≈ 1. 3 that Exceeds the Homogeneity Scale of the R-W Concordance Cosmology.

Monthly Notices of the Royal Astronomical Society, **429,** 2910–2916.

(Large Quasar Group)

8. Urry, C. M., & Padovani, P. (1995).

Unified Schemes for Radio-Loud Active Galactic Nuclei.

Publications of the Astronomical Society of the Pacific, **107,** 803–845.

(Blazars)

Rogue and Runaway Black Holes

9. Sahu, K. C., et al. (2022).

An Isolated Stellar-Mass Black Hole Detected Through Astrometric Microlensing.

The Astrophysical Journal, **933,** 83.

(OB110462)

10. Van Dokkum, P., et al. (2025).

JWST Confirmation of a Runaway Supermassive Black Hole via Its Supersonic Bow Shock.

The Astrophysical Journal Letters.

11. Yao, Y., Chornock, R., Ward, C., et al. (2025).

A Massive Black Hole Revealed by the Offset Tidal Disruption Event AT2024tvd.

The Astrophysical Journal Letters, **985,** L48.

12. Chiaberge, M., et al. (2017).

A Recoiling Supermassive Black Hole in the Quasar 3C 186.

Astronomy & Astrophysics, **600**, A57.

Extreme Physical Limits and Theoretical Concepts

13. Eddington, A.S. (1926).

The Internal Constitution of the Stars.

Cambridge University Press.

(Eddington limit)

14. Shapiro, S.L., & Teukolsky, S. A. (1983).

Black Holes, White Dwarfs, and Neutron Stars: The Physics of Compact Objects.

Wiley.

15. Croker, K.S., et al. (2022).

Cosmologically Coupled Compact Objects: A Single-Parameter Model for Dark Energy.

The Astrophysical Journal, **933,** 214.

16. Farrah, D., et al. (2023).

Observational Evidence for Cosmological Coupling of Black Holes.

The Astrophysical Journal Letters, **944**, L31.

17. Novikov, I.D., & Frolov, V.P. (1989).

Physics of Black Holes.

Springer.

(White holes and theoretical limits)

18. Shapiro, S.L., & Teukolsky, S.A. (1983).

Black Holes, White Dwarfs, and Neutron Stars: The Physics of Compact Objects.

Wiley

Chapter 6 – Physical Characteristics and Extreme Gravitational Effects

1. Misner, C.W., Thorne, K.S., & Wheeler, J.A. (1973).

Gravitation.

W. H. Freeman(publishing company)

2. Frolov, V. P., & Novikov, I.D. (1998).

Black Hole Physics: Basic Concepts and New Developments.

Springer.

3. Shapiro, S. L., & Teukolsky, S.A. (1983).

Black Holes, White Dwarfs, and Neutron Stars: The Physics of Compact Objects.

Wiley.

4. Kerr, R.P. (1963).

Gravitational Field of a Spinning Mass as an Example of Algebraically Special Metrics.

Physical Review Letters, **11,** 237–238.

5. Lense, J., & Thirring, H. (1918).

On the Influence of the Proper Rotation of Central Bodies on the Motion of Planets and Moons.

Physikalische Zeitschrift, **19**, 156–163.

(Frame-dragging effect.)

6. Bardeen, J.M., Carter, B., & Hawking, S.W. (1973).

The Four Laws of Black Hole Mechanics.

Communications in Mathematical Physics, **31,** 161–170.

7. Bekenstein, J.D. (1973).

Black Holes and Entropy.

Physical Review D, **7,** 2333–2346.

8. Hawking, S.W. (1975).

Particle Creation by Black Holes.

Communications in Mathematical Physics, **43,** 199–220.

9. Schneider, P., Ehlers, J., & Falco, E.E. (1992).

Gravitational Lenses.

Springer.

10. Einstein, A. (1916).

The Foundation of the General Theory of Relativity.

Annalen der Physik, **49**, 769–822.

11. Blandford, R.D., & Znajek, R.L. (1977).

Electromagnetic Extraction of Energy from Kerr Black Holes.

Monthly Notices of the Royal Astronomical Society, **179,** 433–456.

12. Krolik, J.H. (1999).

Active Galactic Nuclei: From the Central Black Hole to the Galactic Environment.

Princeton University Press.

Chapter 7 – Extraordinary Black Holes and Their Systems

1. McDonald, M., et al. (2012).

A Massive, Cooling-Flow-Induced Starburst in the Core of the Phoenix Cluster.

Nature, **488,** 349–352.

2. McDonald, M., et al. (2015).

The Remarkable Core of the Phoenix Cluster.

*The Astrophysical Journal, **811,** 111.*

3. Russell, H.R., McDonald, M., McNamara, B.R., et al. (2017).

ALMA Observations of Molecular Gas Filaments in the Phoenix Cluster.

*The Astrophysical Journal. **4***

Extremely Massive Quasars (TON 618)

4. Shen, Y., Richards, G.T., Strauss, M.A., et al. (2011).

A Catalog of Quasar Properties from Sloan Digital Sky Survey Data Release 7.

*The Astrophysical Journal Supplement Series, **194,** 45.*

5. Traktenbrot, B., & Netzer, H. (2012).

Black Hole Mass Estimates and Rapid Growth of Supermassive Black Holes in Luminous Quasars.

The Astrophysical Journal.

6. Volonteri, M. (2010).

Formation of Supermassive Black Holes.

*Astronomy and Astrophysics Review, **18,** 279–315.*

Earliest and Most Distant Black Holes

7. Bogdán, Á., Goulding, A.D., Natarajan, P., et al. (2023).

Evidence for Heavy-Seed Black Holes in the Early Universe from JWST and Chandra Observations of UHZ1.

Nature Astronomy.

8. Goulding, A.D., Bogdán, Á., Natarajan, P., et al. (2023).
JWST Observations of a Highly Redshifted Black Hole Candidate in the Galaxy UHZ1.

Nature Astronomy.

9. Matthee, J., et al. (2024).
CAPERS-LRD-z9: A Candidate Massive Black Hole in a Little Red Dot Galaxy at $z \approx 9$.

Nature Astronomy.

Black Hole Spin Measurements

10. Risaliti, G., Harrison, F.A., Madsen, K.K., et al. (2013).
A Rapidly Spinning Supermassive Black Hole at the Centre of NGC 1365.

*Nature, **494**, 449–451.*

11. Walton, D.J., et al. (2014).
The High-Spin Black Hole in NGC 1365 Observed with NuSTAR and XMM-Newton.

*The Astrophysical Journal, **788**, 76.*

12. Gou, L., McClintock, J.E., Steiner, J.F., et al. (2010).
Estimating the Spin of Stellar-Mass Black Holes via the Continuum-Fitting Method: The Case of A0620-00.

*The Astrophysical Journal, **718**, L122–L126.*

13. Cantrell, A.G., Bailyn, C.D., McClintock, J.E., et al. (2010).
The Inclination, Distance, and Mass of the Black Hole in A0620-00.

*The Astrophysical Journal, **710**, 1127–1141.*

14. McClintock, J.E., & Remillard, R.A. (2006).
Black Hole Binaries.

In Compact Stellar X-Ray Sources. Cambridge University Press.

(Includes GRS 1915+105 high-spin measurements.)

Nearby Black Hole Discoveries

15. El-Badry, K., Rix, H. -W., Quataert, E., et al. (2023).
A Sun-Like Star Orbiting a Black Hole.

Monthly Notices of the Royal Astronomical Society.

(Discovery of Gaia BH1, the nearest known black hole.)

16. Nagarajan, P., El-Badry, K., Triaud, A.H.M.J., et al. (2023).
ESPRESSO Observations of Gaia BH1: High-Precision Orbital Constraints.

arXiv preprint.

Brightest Known Quasar

17. Wolf, C., Onken, C.A., Lai, S., et al. (2024).
The Most Luminous Known Quasar Powered by a 17-Billion-Solar-Mass Black Hole.

Nature Astronomy.

18. Wolf, C., Onken, C.A., et al. (2024).
The Accretion of a Solar Mass per Day by a 17-Billion-Solar-Mass Black Hole.

Nature Astronomy.

Young Black Hole Candidate (W49B)

19. Lopez, L.A., et al. (2013).
Evidence for a Jet-Driven Supernova in W49B.

The Astrophysical Journal.

20. Zhou, P., & Vink, J. (2018).
Asymmetric Type-Ia Supernova Origin of W49B Revealed by Spatially Resolved X-Ray Spectroscopy.

Astronomy & Astrophysics, ***615,*** *A150.*

Extreme Jet Systems

21. Oei, M.S.S.L. , et al. (2024).
The Giant Radio Galaxy Porphyrion: One of the Largest Black Hole Jet Systems Known.

Nature Astronomy.

22. Oei, M.S.S.L., et al. (2023).
The Discovery of Extremely Large Giant Radio Galaxies in LOFAR Surveys.

Astronomy & Astrophysics, ***675,*** *A2.*

23. Schwartz, D.A., Siemiginowska, A., Snios, B., et al. (2020).
Two Candidate High-Redshift X-Ray Jets Without Coincident Radio Jets.

The Astrophysical Journal.

24. Snios, B., Schwartz, D.A., Siemiginowska, A., et al. (2022).
X-Ray Jets in High-Redshift Quasars Including J1610+1811.

The Astrophysical Journal.

Close Pairs of Supermassive Black Holes

25. Voggel, K.T., Hilker, M., Richtler, T., et al. (2022).
Evidence for a Close Pair of Supermassive Black Holes in the Galaxy NGC 7727.

Astronomy & Astrophysics.

26. Kormendy, J., & Ho, L.C. (2013).
Coevolution (or Not) of Supermassive Black Holes and Host Galaxies.

*Annual Review of Astronomy and Astrophysics, **51,** 511–653.*

27. Adams, S. M., Kochanek, C.S., et al. (2017).
The Search for Failed Supernovae with the Large Binocular Telescope.

Monthly Notices of the Royal Astronomical Society.

(Direct collapse of a massive star into a black hole.)

28. Onoue, M., et al. (2026).
Rapid Growth of a Supermassive Black Hole Revealed by Subaru Telescope Observations.

Publications of the Astronomical Society of Japan / Subaru Telescope Research Report.

Chapter 8: The Growth, Fate, and Significance of Black Holes

1. Alexander, T., & Hickox, R.C. (2012).

What Drives the Growth of Black Holes?

*New Astronomy Reviews, 56, **93–121.***

2. Shapiro, S.L., & Teukolsky, S.A. (1983).

Black Holes, White Dwarfs, and Neutron Stars.

Wiley.

3. Volonteri, M. (2012).

The Formation and Growth of Massive Black Holes.

Science, 337, 544–547.

4. Soltan, A. (1982).

Masses of quasars.

Monthly Notices of the Royal Astronomical Society, 200, 115–122.

5. Stephen Hawking (1975).

Particle Creation by Black Holes. Communications in Mathematical Physics, 43, 199–220.

6. Roger Penrose (1969).

Gravitational Collapse: The Role of General Relativity. Rivista del Nuovo Cimento.

7. Roger Blandford & Roman Znajek (1977).

Electromagnetic Extraction of Energy from Kerr Black Holes.

Monthly Notices of the Royal Astronomical Society, 179, 433–456.

8. LIGO Scientific Collaboration and Virgo Collaboration (2016).

Observation of Gravitational Waves from Binary Black Hole Merger. Physical Review Letters, 116, 061102.

9. John Kormendy & Luis Ho (2013).

Coevolution (Or Not) of Supermassive Black Holes and Host Galaxies.

Annual Review of Astronomy and Astrophysics, 51, 511–653.

10. Kip Thorne (1994).

Black Holes and Time Warps: Einstein's Outrageous Legacy. W. W. Norton.

11. Valeri Frolov & Andrei Zelnikov (2011).

Introduction to Black Hole Physics. Oxford University Press.

12. NASA / ESA.

Review articles on black holes

www.ingramcontent.com/pod-product-compliance
Lightning Source LLC
Chambersburg PA
CBHW041748220926
51847CB00013B/139
9789374265765